The Dinosaurs
and their
Mysterious Demise

By the same author
(www.andrew-norman.co.uk)

By Swords Divided: Corfe Castle in the Civil War (Halsgrove, 2003)
Thomas Hardy: Christmas Carollings (Halsgrove, 2005)
Enid Blyton and her Enchantment with Dorset (Halsgrove, 2005)
Tyneham: A Tribute (Halsgrove, 2007)
Agatha Christie: The Finished Portrait (Tempus, 2007)
The Story of George Loveless and the Tolpuddle Martyrs (Halsgrove, 2008)
Father of the Blind: A Portrait of Sir Arthur Pearson (The History Press, 2009)
Agatha Christie: The Pitkin Guide (Pitkin Publishing, 2009)
Arthur Conan Doyle: The Man behind Sherlock Holmes (The History Press, 2009)
HMS Hood: Pride of the Royal Navy (The History Press, 2009)
Purbeck Personalities (Halsgrove, 2009)
Bournemouth's Founders and Famous Visitors (The History Press, 2010)
Thomas Hardy: Behind the Mask (The History Press, 2011)
A Brummie Boy goes to War (Halsgrove, 2011)
Winston Churchill: Portrait of an Unquiet Mind (Pen & Sword Books, 2012)
Charles Darwin: Destroyer of Myths (Pen & Sword Books, 2013)
Beatrix Potter: Her Inner World (Pen & Sword Books, 2013)
T.E. Lawrence: Tormented Hero (Fonthill, 2014)
Agatha Christie: The Disappearing Novelist (Fonthill, 2014)
Lawrence of Arabia's Clouds Hill (Halsgrove, 2014)
Jane Austen: Love is Like a Rose (Fonthill, 2015)
Kindly Light: The Story of Blind Veterans UK (Fonthill, 2015)
Thomas Hardy at Max Gate: The Latter Years (Halsgrove, 2016)
Corfe Remembered (Halsgrove, 2017)
Thomas Hardy: Bockhampton and Beyond (Halsgrove, 2017)
Mugabe: Monarch of Blood and Tears (Austin Macauley, 2017)
Making Sense of Marilyn (Fonthill, 2018)
Hitler's Insanity: A Conspiracy of Silence (Fonthill, 2018)
The Unwitting Fundamentalist (Austin Macauley, 2018)
Robert Mugabe's Lost Jewel of Africa (Fonthill, 2018)
Hitler: Dictator or Puppet? (Pen & Sword, 2011, 2020)
Halsewell: A Shipwreck that Gripped the Nation (Fonthill, 2020)

The Dinosaurs and their Mysterious Demise

Andrew Norman

WHITE OWL

AN IMPRINT OF PEN & SWORD BOOKS LTD
YORKSHIRE – PHILADELPHIA

First published in Great Britain in 2024 by
WHITE OWL
an imprint of Pen & Sword Books Ltd
Yorkshire – Philadelphia

Copyright © Andrew Norman, 2024

ISBN 978-1-39904-112-6

The right of Andrew Norman to be identified as the author of this work has been asserted by him in accordance with the Copyright, Designs and Patents Act 1988.

A CIP catalogue record for this book is available from the British Library.

All rights reserved. No part of this book may be reproduced or transmitted in any form or by any means, electronic or mechanical including photocopying, recording or by any information storage and retrieval system, without permission from the Publisher in writing.

Typeset by Concept, Huddersfield, West Yorkshire, HD4 5JL.
Printed and bound in England by CPI Group (UK) Ltd, Croydon, CR0 4YY.

Pen & Sword Books Ltd incorporates the imprints of Aviation, Atlas, Family History, Fiction, Maritime, Military, Discovery, Politics, History, Archaeology, Select, Wharncliffe Local History, Wharncliffe True Crime, Military Classics, Wharncliffe Transport, Leo Cooper, The Praetorian Press, Remember When, White Owl, Seaforth Publishing and Frontline Books.

For a complete list of Pen & Sword titles please contact
PEN & SWORD BOOKS LTD
47 Church Street, Barnsley, South Yorkshire, S70 2AS, England
E-mail: enquiries@pen-and-sword.co.uk
Website: www.pen-and-sword.co.uk
or
PEN & SWORD BOOKS
1950 Lawrence Rd, Havertown, PA 19083, USA
E-mail: uspen-and-sword@casematepublishers.com
Website: www.penandswordbooks.com

Contents

List of Plates vii
Acknowledgements viii
Introduction ix
1. The Coastal Town of Swanage, Dorset 1
2. Swanage: Ammonites and Evidence of Other Ancient Forms of Life 3
3. The 'Jurassic Coast' 5
4. Purbeck: Strata in Which Dinosaur Fossils are Most Likely to be Found? 8
5. What is a Dinosaur? 13
6. The Crystal Palace Dinosaurs: Sir Arthur Conan Doyle and *The Lost World* 19
7. Mary Anning: Fossil Hunter Extraordinaire! 29
8. Dinosaur Prints: How to Identify Them 32
9. Dinosaur Prints: Some Local Discoveries 37
10. Some Dinosaur Fossils from Purbeck and Further Afield 42
11. How is the Age of the Fossilised Bone of a Dinosaur Ascertained? 47
12. How Life Began, How Living Creatures Evolved, and the Dinosaurs in Particular 48
13. The KT Boundary and the Great KT Extinction Event 51
14. The Presence of High Concentrations of Iridium at the KT Boundary, and its Significance 55
15. Chicxulub: A Possible Location for the Alleged Asteroid Impact 58
16. Was There More Than One Asteroid Impact? Shiva 62
17. Oil: Another Potentially Lethal Ingredient 65

18. The Volcanoes of India's Deccan Region	68
19. The Chicxulub Impact: Both a Local and a Global Catastrophe	70
20. Some Terrestrial, Semi-Aquatic and Marine Creatures that Survived the KT Extinction Event	79
21. Were Dinosaurs Warm-Blooded?	85
22. The Ability to Hibernate or Brumate: The Critical Factor	86
23. Hibernation: Some of the Complexities Involved	93
24. The Genetics of Hibernation	97
25. For the Mammals Shall Inherit the Earth!	99
26. A New Dinosaur is Discovered	101
27. Birds: The Dinosaurs that Did Not Die	105
28. How Did Birds Survive the KT Extinction Event?	109
29. The Enduring Attraction of Dinosaurs	112
Epilogue	114
Appendix: Dinosaur Data	119
Notes	131
Bibliography	145
Index	147

List of Plates

Isle of Purbeck.
Swanage's guardian dinosaur!
Ammonite built into the wall of the Tithe Barn, Swanage.
William Buckland. Mezzotint by S. Cousins, 1833, after T. Phillips.
Benjamin Waterhouse Hawkins at work on the iguanodon mould in his model room at Crystal Palace Park, Sydenham Hill.
Illustration by Patrick L. Forbes for Sir Arthur Conan Doyle's novel *The Lost World*.
Mary Anning by Benjamin Donne.
Mary Anning's Commonplace Book.
Henry de la Beche.
Glen J. Kuban on the Ozark Site, Dinosaur Valley State Park, Texas.
Dinosaur tail vertebrae discovered at Peveril Point, Swanage.
Iguanodon tail vertebra discovered in Wealden beds.
Cast of dinosaur hind-foot print, probably that of an *Iguanodon*, found near Swanage.
Jawbone of dinosaur *Nuthetes*.
The Common Poorwill (*Phalaenoptilus nuttalii*).
The four spinal fossilized vertebrae of *Vectaerovenator inopinatus*.
Dinosaur vertebra discovered by Paul I. Farrell on Shanklin beach.
Dinosaur Isle Museum, Sandown, Isle of Wight.
Life-sized cast of the sauropod dinosaur *Patagotitan* from Patagonia, Argentina.
Patagotitan, 55–63 tons in weight and 102 feet in length, tail extending into an adjacent room!
Patagotitan: fossilised 7 feet 10 inch long femur (thigh bone).
Patagotitan, with three dinosaur enthusiasts.

Acknowledgements

I am especially grateful to Charles G. Bardeen, Dougal Dixon, Martin Munt, and Glen J. Kuban. I am also grateful to the following: Darin Ball; Peter Bass; Izzy Farrell; Paul I. Farrell; Gordon C. Grigg; David Haysom; Kevin W.H. Keates (senior); Kevin A. Keates (junior); Barry L. King; Lori Little; Ivor McNeill; Robert J. Pugh; Ella Reece; Jo, Darcy, and Henry Tapper; Diana and Alan Turner; Joanna Wright; and Noel O'Neill.

My thanks extend to The British Library; Dorset County Museum, High West Street, Dorchester, Dorset; The Etches Collection, Kimmeridge, Dorset; and Swanage Museum and Heritage Centre, The Square, Swanage, Dorset.

I am especially grateful to my beloved wife, Rachel, for all her help and encouragement.

Introduction

An entirely new theory as to why the dinosaurs failed to fulfil Charles Darwin's maxim, 'the survival of the fittest', and why they exist today only as fossils and not as living creatures, having become totally extinct, is proposed in this volume.

The dinosaurs are a source of endless fascination, and each new generation is inspired and enchanted by images of these wondrous and awe-inspiring creatures that dominated the Earth eons ago. The smallest was the size of a chicken; the largest on record, the titanosaur *Argentinosaurus huinculensis*, weighed about 95 tons – fifteen times as much as an African bull elephant (today's largest terrestrial creature).

Equally intriguing is why the dinosaurs disappeared, having ruled the Earth for no less than 181 million years: about 600 times longer than *Homo sapiens* have existed on the planet. (*Homo sapiens* is defined as the primate species to which modern humans belong: the first modern humans having evolved in Africa about 300,000 years ago. Primate: mammal of an order (Primates) that includes monkeys, apes and humans.[1])

Various theories have been put forward to explain this – some more plausible than others. For example, in 1982, British ophthalmologist Laurence R. Croft suggested that the reason for the dinosaurs' extinction was that the brightness of the sun induced cataracts that made them go blind before they reached sexual maturity (i.e. they were unable to breed on this account).[2]

Recently, an anonymous blogger on the Internet humorously echoed this theory by stating that 'over exposure to sun & UV [ultraviolet light] caused cataract blindness that led dinos not to be able to see & thus fall off cliffs to their extinction. Yeah Ok!' A theory as to what actually befell the dinosaurs will be put forward shortly.

* * *

Swanage is a seaside resort on the south coast of England in the county of Dorsetshire (Dorset). Floating up and down on the thermal currents of air, alongside the cliffs to the south of the town, between Durlston Head and St Aldhelm's Head, kittiwakes are to be seen. (Kittiwake: a small gull that nests in colonies on sea cliffs and has a loud call resembling its name.[3]) They spend most of their time at sea and return to breed on these south-facing cliffs, where large colonies of them congregate.

As evening falls, a solitary cormorant, skimming a foot or two above the waves, makes its way from Poole Harbour across Poole Bay to the Old Harry Rocks, on the north side of Swanage Bay. On these steep chalk cliffs, safe from predators, the cormorants spend the night. Both these species of bird, and all other birds worldwide, strictly speaking, should not be here at all. In fact, they should all have become extinct, intuitively speaking, at the end of the Cretaceous Period 66 million years ago (mya). The reason for this, and the explanation for their survival, will be discussed shortly.

Chapter 1

The Coastal Town of Swanage, Dorset

The town of Swanage is one with which the author is highly familiar, for he is a resident of the English county of Dorsetshire ('Dorset') in which the coastal resort is located.

Swanage lies in the 'Isle of Purbeck' (or 'Purbeck'), which is, in fact, not an island at all, but a peninsula of land. Purbeck is bordered to the south by the English Channel; to the east by Durlston Bay, Swanage Bay and Studland Bay; and to the north by Poole Harbour. It comprises an area of some 60 square miles.

The town of Swanage lies towards the southern end of Swanage Bay and is bounded on the north side by Ballard Down, which towers 400 feet above sea level, and at the southern end by Peveril Point, beyond which a jagged ledge of rocks projects into the sea, waiting to ensnare the unwary sailor. The town itself faces the Isle of Wight – a true island 17 miles or so to the east.

At the time of the compilation of the *Domesday Book* (a comprehensive inventory of the extent, value and ownership of land in England, made in the year 1086 by order of King William I), Swanage was known as 'Sonwic' or 'Swanwic'. At that time, the landowners were recorded as being the wife of Hugh Fitz Grip, and the Countess of Boulogne.[1]

It is no coincidence that the older properties in Swanage are almost invariably constructed of stone, which is exceedingly plentiful thereabouts. In one such stone-built house, in Swanage High Street, lived Petty Officer Edgar Evans (1876–1912), a Welsh naval officer and member of Captain Robert Falcon Scott's ill-fated *Terra Nova* expedition to the South Pole in 1911–12. Evans was one of the group of five selected for the final push to the Pole – all of whom perished on the return journey to base camp.

In recent times, Swanage is where the famous children's author Enid Blyton stayed for her holidays, and it is in the Isle of Purbeck that she set the scene for *The Famous Five* adventure stories featuring Julian, Dick, Anne, Georgina and their dog, Timmy.

The Dinosaurs and their Mysterious Demise

In 2018, Enid's characters featured in the German film *The Famous Five in the Valley of the Dinosaurs*, directed by Mike Marzuk.

Now, the author himself is about to embark on a great adventure of his own; in search not of villains, such as those who featured in such rippingly good tales as *Five Go to Smuggler's Top* (1945), but of truly remarkable creatures that lived in this vicinity an unimaginably long time ago. For Swanage is at the eastern end of the 'Jurassic Coast', which is literally a mine of information in respect of the current subject in question: the dinosaurs. (Jurassic Period: 201.3–145 million years ago (mya), which was the second period of the Mesozoic Era – i.e. between the earlier Triassic Period and the later Cretaceous Period.)

Chapter 2

Swanage: Ammonites and Evidence of Other Ancient Forms of Life

Those who are familiar with Swanage will be aware of the ancient tithe barn, at the lower end of Church Hill and on the opposite side of the road to the Parish Church of St Mary the Virgin. They will also be aware of a large, fossilised ammonite built into its wall.

What is a fossil? A fossil is defined as the remains or impression of a prehistoric plant or animal embedded in rock and preserved in petrified form.[1] Fossils are commonly created by permineralisation. (Permineralisation: fossilisation through the precipitation of dissolved minerals in the interstices of hard tissue. Interstice: small intervening space.[2])

When a creature such as a deceased dinosaur became buried, the parts of that creature that were filled with liquid or gas during its lifetime 'become filled with mineral-rich groundwater. Minerals precipitate from the groundwater, occupying the empty spaces. This process can occur in very small spaces [such as within the walls of the cells]. Small-scale permineralisation can produce very detailed fossils.'

For permineralisation to occur, the organism must become covered by sediment soon after death or soon after the initial decay process. The degree to which the remains are decayed when covered determines the later details of the fossil. Some fossils consist only of skeletal remains or teeth; other fossils contain traces of skin, feathers, or even soft tissues.[3]

This process is known as 'diagenesis', referring to the physical and chemical changes occurring during the conversion of sediment to sedimentary rock.[4]

Ammonites are an extinct group of squid-like marine molluscs of the class Cephalopoda, subclass Ammonoidea, with soft bodies and a hard shell. The shell of the ammonoid was spiral-shaped and comprised numerous gas-filled chambers; the outermost of which contained the body of the creature. (Marine: relating to, or found, in the sea.[5])

Ammonites first appeared in the Devonian Period (419.2–358.9 mya) of the Palaeozoic Era, and disappeared with the KT extinction event. (The

KT extinction: a sudden mas extinction of some three-quarters of all plant and animal species on Earth, approximately 66 mya.)

The ammonite is named after the Amon, an ancient deity who was revered as king of the Egyptian gods (*circa* twenty-first century BC). This is because Amon is sometimes represented as a ram's head or as a ram, the horns of which resemble the ammonite's spiral shells.

This particular specimen, which measures 28 inches by 32 inches, is probably that of a female: the larger size reflecting the fact that it is the females that carry and lay the eggs. The males are smaller and have horns or lappets. (Lappet: a fold or hanging piece of flesh.[6])

Furthermore, many of Swanage's older, stone-built properties also have ammonites built into their front walls and are therefore on display to the passerby. These have been, and continue to be, excavated by local quarrymen during their labours. Many more extraordinary creatures are to be found in the rocks and geological strata of southern England's Jurassic Coast, as will shortly be seen. (Stratum: a layer or a series of layers of rock.[7])

Chapter 3

The Jurassic Coast

The Jurassic Coast in Southern England is a UNESCO World Heritage Site. It stretches from Studland Bay in Purbeck, Dorset, in the east, to Exmouth, Devonshire, in the west – a distance of about 96 miles. The site contains an almost uninterrupted rock record of the Mesozoic Era (251.9–66 mya).

Geologists have divided the Earth's 4,500 million years of history into:

- eons (a major division of geological time[1]);
- eras (a subdivision of an eon);
- epochs (a subdivision of an era); and
- periods (a subdivision of an epoch).

The Phanerozoic (from 541 mya) denotes an eon covering the whole of time since the beginning of the Cambrian Period to the present day. The middle of the three eras that make up the Phanerozoic is the Mesozoic Era (251.9–66 mya), and it was during this era that the dinosaurs lived. The Mesozoic Era comprises the Triassic, Jurassic and Cretaceous Periods.

The Triassic Period (251.9–201.3 mya)

The Earth's land mass during the Triassic Period consisted of a single supercontinent, given the name 'Pangea'. At that time, the Jurassic Coast lay just north of the Equator, in the centre of Pangea.

During this period, 'the climate was relatively hot and dry, and much of the land was covered with large deserts. Unlike today, there were no polar ice caps. It was in this environment that reptiles known as dinosaurs first evolved', with the first dinosaurs appearing about 247 mya.[2,3]

Towards the end of the Triassic Period, a series of earthquakes and massive volcanic eruptions caused Pangaea slowly to begin to break into two. This was the birth of the North Atlantic Ocean.

At the end of the Triassic Period there was a mass extinction, the causes of which are still hotly debated. Many large land animals were wiped out, but the dinosaurs survived, giving them the opportunity to evolve into a wide variety of forms and increase in number.[4]

The Jurassic Period (201.3–145 mya)

The single land mass, Pangaea, split into two, creating [the ancient continents of] Laurasia in the north and Gondwana in the south. Despite this separation, similarities in their fossil records show that there were some land connections between the two continents early in the Jurassic. These regions became more distinct later in the period.

To the east, Tethys [the ocean between Gondwana and Laurasia] is actively spreading, while the British Isles continue to drift northwards to 20' to 30' north, equivalent to the latitude of the modern-day Sahara.[5]

The break-up of Pangaea results in Gondwana and Laurasia separating, as the southern Atlantic Ocean starts to rift [crack or split] open. The British Isles continue to drift northwards to 35' to 40' north into more temperate conditions, with lithospheric extension [lithosphere: the rigid outer part of the earth, consisting of the crust and upper mantle] and passive rifting occurring to the east (forming the North Sea) and west (where later the North Atlantic [Ocean] will open)."[6]

Temperatures fell slightly, although it was still warmer than today due to higher amounts of carbon dioxide in the atmosphere. Rainfall increased as a result of the large seas appearing between the land masses.

These changes allowed plants such as ferns and horsetails to grow over huge areas. Some of this vegetation became the fossil fuels that we mine today. Elsewhere there were forests of tall conifer trees such as sequoias and monkey puzzles. The plentiful plant supply allowed the huge plant-eating sauropods to evolve.

These are some of the largest animals ever to have walked the Earth. By the end of the Jurassic their herds dominated the landscape. Sauropods became even larger in the Cretaceous.[7] (Sauropod: an order of huge herbivorous dinosaurs with long necks and tails and massive limbs.[8])

The Cretaceous Period (145–66 mya)

During the Cretaceous the land separated further into some of the continents we recognise today, although in different positions. This meant that dinosaurs evolved independently in different parts of the world, becoming more diverse.

Other groups of organisms also diversified. The first snakes evolved during this time, as well as the first flowering plants. Various insect groups appeared, including bees, which helped increase the spread of flowering plants. And mammals now included tree climbers, ground dwellers, and even predators of small dinosaurs.

Sea levels rose and fell repeatedly during the Cretaceous Period. At the highest point there were many shallow seas separating parts of the continents we know today. For example, Europe was made up of many small islands. Thick layers of sediment built up at the bottom of these seas as single-celled algae died and their skeletons fell to the seabed.[9]

This is how chalk was first formed. In fact, the word 'Cretaceous' derives from *creta*, the Latin word for chalk.

By the Late Cretaceous–Early Palaeogene (first period of the Cenozoic Era, following the Cretaceous), 'passive rifting has given way to active rifting to the west of the British Isles, allowing the northern Atlantic Ocean to continue opening in zipper-like fashion, northwards. Active sea floor spreading is occurring throughout the Atlantic, Indian, and Pacific Oceans. Whilst [the ocean] Tethys closes, resulting in the eventual collision of Africa, India, and Eurasia.'[10]

With the break-up of Pangea into separate continents, members of the global dinosaur family became separated. This led to greater diversity, as they evolved along separate lines and adapted to the various environments in which they found themselves. Because the tenure of the dinosaurs on Earth spanned 181 million years or so, this diversity was great indeed.

Chapter 4

Purbeck: Strata in which Dinosaur Fossils are Most Likely to be Found?

It would be straightforward for the fossil hunter if, during the evolution of the Earth, the most recently formed rocks or geological formations had remained at the surface and the most ancient ones at the greatest depth – that is, if the most recent of the Cretaceous Period's rocks were on top, followed at a greater depth by those of the older Jurassic and Triassic Periods respectively. However, life is not as simple as that. As Swanage historians David Lewer and Dennis Smale pointed out, over hundreds of millions of years, Purbeck has successively 'sunk beneath the waves, [been] raised up again, convulsed, tilted, denuded, flooded, sun-baked, and frozen, for time out of mind'.[1]

Geologist Robert Damon's geological map of Purbeck (1884) illustrates the location of the various geological formations to be found in the Isle of Purbeck.[2] This shows that in this locality, due to the aforementioned upheavals in the Earth's crust, the oldest rocks are to be found on the southern coastline; the next oldest further inland, and so forth.

How thick were these geological formations, when originally laid down? Because these layers can change dramatically, not only when the deposited strata are uplifted, distorted or eroded but also when they are compacted, their thickness may be difficult to elucidate. However, the British Geological Survey (BGS) has produced charts showing the approximate values.[3]

Imagine a journey northwards, commencing at Dancing Ledge, the site of a former coastal cliff quarry 2½ miles or so to the south-west of Swanage. (Note: A particular stratum or layer of rock is known as a 'Group', or alternatively as a 'Formation' or a 'Bed'; for example, the 'Portland Limestone Group', or simply the 'Portland Group'. Group: a lithostratigraphic unit, part of the geologic record that consists of defined rock strata: may sometimes be divided into subgroups.[4]) In this case, the geological strata encountered follows, travelling from south to north.

The Portland Group

- Limestone, sedimentary marine deposit.
- Estimated original thickness: 100 feet (BGS).
- Created: 147.6–144.8 mya, during the final age of the Late Jurassic and the first age of the Early Cretaceous.[6]

This is a sedimentary oolitic limestone, off-white or cream in colour. (Oolite: limestone consisting of a mass of rounded grains made up of concentric layers.[7]) The grains ('ooliths') are created when a fragment of shell or sand acts as a 'seed', around which layers or calcium carbonate precipitate from supersaturated water.

It should be noted that those strata, which are exclusively marine (relating to or produced by the sea) in origin, do not normally contain dinosaur imprints or remains; these being terrestrial creatures.[8]

The Purbeck Group

- Limestone, sedimentary deposit, partly marine and partly freshwater or brackish lagoonal.
- Estimated original thickness: 300 feet (BGS).
- Created: 144.8–136.8 mya, during the first age of the Early Cretaceous.

The rock strata of the Purbeck Group, which overlies that of the Portland Group, is subdivided into Lower (Early) Purbeck, Middle Purbeck and Upper (Late) Purbeck Beds. (Brackish: slightly salty, as in river estuaries. Lagoon: a stretch of salt water separated from the sea by a low sandbank; a small freshwater lake near a larger lake or river.[9])

Continuing northwards, the traveller now traverses the Purbeck Beds. This sedimentary limestone often has a very high fossil content and varies between pale and dark grey, sometimes with hints of yellow or pink.

Palaeontologist Robert A. Coram stated:

> About 135 million years ago, Dorset underwent a great change of face. For many millions of years, the county had been part of the seabed [until] eventually, forces deep within the Earth buckled the sea floor and forced it upwards, above sea level, to form a giant northern land mass. Dorset lay at the southern edge of this land mass: a low-lying, sub-tropical landscape which was periodically submerged by

salty arms of the sea called lagoons, or by huge freshwater lakes. A series of rock layers were deposited which are now called the Purbeck Beds.[10]

Coram continued that if it were possible to travel back in time and visit the Isle of Purbeck 135 million years ago (to the Valanginian Age of the Early Cretaceous Period), you would 'find yourself in the middle of an immense forest':

> You'd see that the trees were mainly of one type, which has now been given the tongue-torturing name of *Protocupressinoxylon purbeckensis*, which was basically an evergreen tree very similar to the modern Cypress. It had tiny scaly leaves and a stout trunk that probably extended to a height of well over 20 metres (60 feet). [The] remains of these trees are quite common in some of the Dirt Beds. [Dirt Bed, or palaeosol: soil preserved by burial under sediments that, in the case of older deposits, has lithified. Lithify: transform into stone.[11]]
>
> At the [so-called] Fossil Forest, east of Lulworth Cove, a large area of what was once the forest floor is now exposed half-way up the cliff face. It is possible to pick up handfuls of the ancient soil and stand between the rock-encrusted stumps and horizontal trunks that lie exactly where they fell, 135 million years ago.

Cycads, with 'squat knobbly trunks from which sprouted huge fern-like leaves', were also to be found in the Purbeck Forest. (Cycad: a palm-like plant of tropical and subtropical regions, bearing large male or female cones.[12])

Also, said Coram, 'insects such as dragonflies, cockroaches and beetles' are also commonly to be found in rocks of 'Purbeck Forest age'. In the lakes and ponds lived 'snails, various types of fishes, and turtles and crocodiles', and 'among the foliage around the ponds darted lizards and primitive, furry mammals'.[13] (Primitive: relating to or denoting the earliest stages in evolution or development.[14])

To summarise, the Purbeck Group consists of sedimentary limestones frequently separated by clays and/or shales that may be more or less calcareous. (Clay: a stiff, sticky, fine-grained impermeable earth. Shale: soft, finely stratified sedimentary rock formed from consolidated mud or clay. Calcareous: containing calcium carbonate: chalky.[15])

Within these beds can be found the remains of the creatures mentioned above, together with, not least, a veritable cornucopia of fossilised

Purbeck: Strata in which Dinosaur Fossils are Most Likely to be Found?

dinosaur remains, including trackways and prints, as will shortly be seen. (Track: a trail of prints.)

(Nomenclature: In this narrative, the impression made by the dinosaur's forefoot is referred to as the forefoot print and the impression made by the dinosaur's hind foot is referred to as the hind-foot print. The infill is called the 'natural cast'. The digits of the forefoot are referred to as fingers, and those of the hind foot as toes.)

The Purbeck Marble Bed

- Sedimentary deposit; freshwater.
- Estimated original thickness: not more than 4 feet or so.
- Created: during the first age of the Early Cretaceous.

The thin seam of 'Purbeck Marble' is not a true marble but a sedimentary limestone not more than 4 feet or so in thickness. The marble seam runs from Peveril Point, Swanage, westwards to Worbarrow Bay – a distance of about 11 miles. It was laid down in shallow freshwater lagoons, and it dates from the Berriasian Age (145–139.8 mya) of the Early Cretaceous. It is densely packed with fossils of the freshwater snail *Viviparus*.

Wealden Group

- Sedimentary deposit; freshwater.
- Estimated original thickness: up to 1,300 feet (BGS).
- Created: 136.8–124.8 mya, during the Early Cretaceous.

The Wealden Group is now usually referred to as the 'Wessex Formation' where it outcrops in Dorset and the Isle of Wight.

This rock stratum is derived from clay, formed by soil particles, which, in turn, were derived from the weathering of rocks. The material was transported by streams and rivers from higher levels and deposited in braided channels, and also in freshwater lagoons. (Braided: the manner in which a river or stream flows into shallow interconnected channels divided by deposited earth or alluvium.[16])

In the Wealden Beds of Purbeck was discovered a vertebra from the tail of a dinosaur, identified as an *Iguanodon*. This will be discussed shortly.

Lower (Early) Greensand Group, Gault Clay Beds and Upper (Late) Greensand Group

- Sedimentary deposits, marine.
- Estimated original thickness: 230 feet (BGS).
- Created: 124.8–99 mya, during the Early and Mid-Cretaceous.

Greensand is a greenish kind of sandstone, often loosely consolidated. The Gault Clay Beds consist of a series of clays and marls forming strata.[17] (Marl: an unconsolidated sedimentary rock or soil consisting of clay and lime.[18])

Chalk Group

- Sedimentary deposit, marine.
- Estimated original thickness: up to 1,100 feet (BGS).
- Created: 99–66 mya, during the Mid-to-Late Cretaceous.

This brings the traveller to the chalk foothills of the Purbecks. (Chalk: composed of calcite, a white or colourless mineral consisting of calcium carbonate.[19])

Beyond these strata, the London Clay/Reading Beds, and Poole Formation (formerly known as the Bagshot Beds) all date from the post-Cretaceous and, therefore, post-dinosaur period.

Chapter 5

What is a Dinosaur?

In the 1820s, the fossilised bones of several large, lizard-like creatures were discovered in southern England. For example, in 1824, at Stonesfield in Oxfordshire, William Buckland (1784–1856) discovered the bones of one such creature. He named it *'Megalosaurus'* (from the Greek *megas* ('great') and *sauros* ('lizard')). However, this was no ordinary reptile, as will be seen. Buckland was an Anglican clergyman, a Member of Parliament, Professor of Geology at Oxford University, and President of the Geological Society of London.

In 1825, geologist, obstetrician and palaeontologist Gideon Mantell (1790–1852), discovered teeth and, subsequently, an entire skeleton at Whiteman's Green Quarry, Cuckfield, Sussex. As its teeth resembled those of an iguana, he named the creature *'Iguanodon'* (*odont* being Greek for tooth). In 1833, Mantell discovered another species, which he named *Hylaeosaurus* (from the Greek *hylaios* – of the forest). This was later identified as a herbivorous ornithischian ankylosaurian dinosaur.

In 1835, Buckland 'provided the first report of dinosaurs from Dorset', but the term 'dinosaur' had not yet been invented. His report mentioned *Iguanodon* vertebrae and limb elements, and possible material of *Megalosaurus*, from the 'Iron Sand' of Swanage Bay.[1]

In about 1839, biologist, comparative anatomist, palaeontologist and founder of London's Natural History Museum Sir Richard Owen (1804–92) began a study of the fossilised bones of these three, large, lizard-like creatures. Whereupon, he made a remarkable observation: that five vertebrae at the base of their spines were fused together, unlike in the case of other contemporary reptiles. (Reptile: a cold-blooded vertebrate animal of the class Reptilia that includes snakes, lizards, crocodiles, turtles and tortoises, typically having a dry, scaly skin and laying soft-shelled eggs on land. Vertebrate: an animal of a large group, subphylum Vertebrata, distinguished by the possession of a backbone or spinal column, including mammals, birds, reptiles, amphibians and fish.[2])

In 1841, Owen announced his findings in that year's issue of the *Report of the British Association for The Advancement of Science*, entitled 'Report on

British Reptiles'. The creatures also had several other unique skeletal characteristics in common. He said:

> The combination of such characters ... altogether peculiar among Reptiles ... all manifested by creatures far surpassing in size the largest of existing reptiles, will, it is presumed, be deemed sufficient ground for establishing a distinct tribe or suborder of Saurian Reptiles, for which I would propose the name of 'Dinosauria'.

In other words, Owen had coined the term 'dinosaur', from the Greek *deinos* ('terrible') and *sauros* ('lizard').[3] Today, in excess of 300 genera of dinosaurs have been discovered and named.[4]

Early Classification

It was Harry Govier Seeley (1839–1909), a palaeontologist trained in Cambridge under the renowned geologist and priest Adam Sedgwick (1785–1873), who [in 1887] decided that dinosaurs could be divided into two distinct orders, the Saurischia and the Ornithischia. This classification was based on the arrangement of the creatures' hip bones and, in particular, whether they displayed a lizard-like pattern (Saurischia) or a bird-like one (Ornithischia).[5]

The ornithischians and saurischians were at first thought to be unrelated ... but later studies showed that they had all evolved from a single common ancestor.[6] (Ornithischians and saurischians separated as distinct lineages about 235–240 mya, during the Mid-Triassic Period. Ancestor: something from which a later species has evolved.[7])

Megalosaurus is now classified as a carnivorous saurischian theropod dinosaur, and *Iguanodon* is now classified as a herbivorous ornithischian ornithopod dinosaur.

Example: The Classification of *Tyrannosaurus rex*

The taxonomic classification of *Tyrannosaurus rex* is as follows:

- clade – Dinosauria;
- order – Saurischia;
- suborder – Theropoda;
- family – Tyrannosauroidae;
- genus – *Tyrannosaurus*; and
- species – *Tyrannosaurus rex*.

(Taxonomy: the branch of science concerned with classification; a scheme of classification. Taxon: a taxonomic group of any rank. Clade: a group of organisms comprising all the evolutionary descendants of a common ancestor.[8])

The Classification of Dinosaurs
(see Appendix, 'Dinosaur Classification')

Note: Classifications of dinosaurs are subjective and temporary, as new species are discovered and new knowledge gained.

The Saurischia
Dinosaurs of this order fall into two main groups (suborders): the Sauropodomorpha (sauropods) and the Theropoda (theropods).

The Sauropodomorpha
These dinosaurs were quadrupedal and herbivorous. (Quadruped: an animal that has four feet.[9]) They had long necks and tails, proportionally small heads, thick, pillar-like legs and lizard-like feet. They could attain great size. Within the suborder Sauropodomorpha is the clade Titanosauria. This includes the largest land animals ever to have existed, namely members of the genus *Titanosaurus*.

The Theropoda
Theropod dinosaurs were bipedal and carnivorous. (Biped: an animal that walks on two feet.[10] The suborder includes the genus *Tyrannosaurus*, the most fearsome member of which is the mighty *Tyrannosaurus rex*. It also includes the Dromaeosauridae, feathered dinosaurs often called 'raptors'.

The Ornithischia
Dinosaurs of this order were herbivorous. Among their number were those with horns, 'armour plating' in the form of protective plates and scales, domed heads, club-like tails, and duck-bills.

In 2017, the results of a collaborative study by Cambridge University and the Natural History Museum (CU/NHM) were published.[11] The researchers looked at nearly 35,000 individual anatomical characteristic features, 'based on 457 different features found in 74 early dinosaur species. They used a computer programme to model the most likely way that each feature would have evolved and used the results to place the dinosaurs into a new family tree.'

The CU/NHM team concluded that theropods were 'more closely related to ornithischians', and 'that these two groups were indeed part of the same clade'. Accordingly, in their 'revised grouping', Ornithischia and

Theropoda were grouped together as 'Ornithoscelida', meaning 'bird-limbed'.

This classification has, as yet, not been universally adopted.

The Greatest and the Smallest

The largest dinosaur yet discovered, as already mentioned, is the titanosaur *Argentinosaurus huinculensis*, a sauropod herbivore weighing up to 100 tons and with a length of up to 130 feet.

Various dinosaurs compete for the title of the smallest. One is *Compsognathus*, a carnivorous bipedal theropod. About the size of a chicken, it weighed approximately 7 pounds and was about 3 feet in length, when fully grown.

The CU/NHM team stated that it was likely that 'the earliest dinosaurs were relatively small – about 1 to 2 metres long'.

The Oldest Dinosaur Yet Discovered

Although, according to the CU/NHM team, it was a matter of some dispute, 'the earliest animal claimed as a dinosaur' was *Nyasasaurus parringtoni*, 'which lived about 245 million years ago'. If this was indeed the case, it was estimated that this species – i.e. 'the very first dinosaur [had first] appeared around 247 million years ago [i.e. in the Early Triassic Period]'. This, therefore, is when dinosaurs diverged from reptiles, and became a group of their own.

How do Dinosaurs differ from Other Reptiles?

Dinosaurs are classed as reptiles. However, their skeletons have features that distinguish them from other archosaurs. (Archosaur: a reptile of a large group that includes the crocodilians together with the extinct dinosaurs and pterosaurs. Crocodilians: reptiles of the order Crocodylia, comprising the crocodiles, alligators, caimans and gharial. Gharial: a large fish-eating crocodile with a long narrow snout that widens at the nostrils.[12])

Compared with other reptiles, dinosaurs are generally believed to have been warm-blooded. (An ectotherm is defined as an animal that is dependent on external sources of heat to maintain its body temperature.[13] Conversely, an endotherm is an animal that is dependent on the internal generation of heat to maintain its body temperature.[14]) The term 'cold-blooded' is misleading and 'poikilotherm' – an organism that cannot regulate its body temperature except by behavioural means such as basking or burrowing – is more appropriate.[15] (Bask: lie exposed to warmth and sunlight for pleasure. Burrow: a hole or tunnel dug by a small animal as a dwelling.[16])

The Skeleton

Dinosaurs possess a sacrum that is extended by the fusion to it of additional caudal or lumbar vertebrae. (Caudal: at or near the tail.[17])

Whereas, with other reptiles, the head of the thigh bone (femur) fits snugly into the hip socket (acetabulum), dinosaurs possess a perforate hip socket (one with a hole in it), whereby the head of the thigh bone protrudes through an aperture in the hip socket.

'This arrangement allowed the dinosaurs to position their hind legs close to and underneath the body, allowing a full-time erect stance, compared to the more splayed-leg reptilian posture.' Also, the position and orientation of the acetabulum was a morphological trait that enabled the dinosaurs to adopt a more upright posture.

This more upright stance is more efficient for weight-bearing – for economy of energy when in motion, and for maximising speed when running. In particular, it enabled the quadrupeds to stand upright on their hind legs, if the need arose.

The Place of Origin of the Dinosaurs

Although current thinking is that the first dinosaurs appeared 'in the southern hemisphere, which at the time was occupied by the supercontinent Gondwana', the CU/NHM study concluded, 'that the first dinosaurs could also have emerged in the north, on the landmass known as Laurasia'.

Locomotion

The CU/NHM study suggested that the earliest dinosaurs 'walked on two legs', and that these 'primitive dinosaurs could use their front two limbs (no longer needed for walking) as hands to grasp objects. The researchers speculate that these grasping hands may have given early dinosaurs an advantage over rival animals, perhaps in feeding'.

Diet – What Foods did Dinosaurs Consume?

According to the CU/NHM study, the teeth of the early dinosaurs indicated that they 'were probably omnivores', and it was 'therefore likely that the common ancestor of dinosaurs was an omnivore too'. (Omnivorous: feeding on a variety of food, of both plant and animal origin.[18])

Theoretically, by studying coprolites, it should be possible to discover what type of plants the herbivorous dinosaurs ate. (Coprolite: pieces of fossilised dung.[19]) Science writer Jeff Hecht stated:

> Microscopic examination of fossilized dinosaur dung from India now shows that the last massive plant-eating dinosaurs munched heaping helpings of at least five different types of grass. The key evidence is tiny silica crystals called phytoliths which grow inside plant cells and can survive digestion. [These crystals] dated to 65–70 million years old [i.e. they were primarily Late Cretaceous. Furthermore,] the only dinosaur bones found near the fossil dung are those of massive, long-necked plant-eaters called titanosaurs. As for the grasses, they were described as of a 'more highly evolved' type.[20]

US palaeontologist Karen Chin said: 'Here we describe fossilized feces (coprolites) that demonstrate recurring consumptions of crustaceans and rotted wood by large, Late Cretaceous dinosaurs'. The samples came from the Kaiparowits Formation of southern Utah, USA. (Crustacea: a large group of mainly aquatic arthropods, which include crabs, crayfish and lobsters. Arthropods: phylum arthropoda: a large phylum of cold-blooded invertebrate animals having a segmented body, external skeleton and jointed limbs that includes insects, spiders, centipedes, scorpions, mites, ticks, barnacles, crustaceans and their relatives. Phylum: a principal taxonomic category that ranks above class and below kingdom.[21])

The samples of coprolite were 'primarily composed of comminuted [reduced to minute particles or fragments[22]] conifer wood tissues that were fungally degraded before ingestion. Thick fragments of laminar crustacean cuticle [outer shell, or exoskeleton] are scattered within the coprolite content and suggest that the dinosaurian defecators consumed sizeable crustaceans that sheltered in rotting logs.'

This implies that the dinosaurs had swallowed the decomposing wood inadvertently, while rummaging for crabs.[23]

Chapter 6

The Crystal Palace Dinosaurs: Sir Arthur Conan Doyle and *The Lost World*

In the summer of 1854, members of the public were to come face-to-face with dinosaurs in a way that they could not possibly have imagined.

The Great Exhibition (of the Works and Industry of all Nations) was opened by Queen Victoria on 1 May 1851 in London's Hyde Park. Exhibits were in a huge edifice of glass and steel, designed by architect and landscape gardener Joseph Paxton, and known as the 'Crystal Palace'.

The exhibition closed on 11 October 1851, having been visited by about 6 million people. Whereupon, the Crystal Palace was re-erected in the grounds of Penge Place, Sydenham Hill, South London: the palace and site having been purchased by the newly formed Crystal Palace Company. Here, in the grounds of 'Crystal Palace Park', lakes would be created, complete with islands on which thirty-three life-sized concrete sculptures of prehistoric creatures would be placed. This would include four dinosaurs: two *Iguanodons*, one *Hylaeosaur* and one *Megalosaurus*. Not only that, but with the aid of fountains, the water levels would be made successively to rise and fall, whereupon the creatures would be covered by the tide, only to be subsequently revealed, much to the excitement of the onlookers. Each island represented an era of time: firstly the Palaeozoic; secondly the Mesozoic; and finally the Cenozoic.

Benjamin Waterhouse Hawkins (1807–94), artist and sculptor, who was assistant superintendent of the Great Exhibition, built the dinosaurs under the supervision of Sir Richard Owen. His method, in his own words:

> ... to make preliminary drawings, with careful measurements of the fossil bones in our Museum of the College of Surgeons, British Museum, and Geological Society: thus prepared I made my sketch-models to scale, either a 6th or 12th of the natural size, designing such attitudes as my long acquaintance with the recent and living

forms of the animal kingdom enabled me to adapt to the extinct species I was endeavouring to restore. These sketch-models I submitted in all instances to the criticism of Professor Owen, who with his great knowledge and profound learning most liberally aided me in every difficulty. His sanction and approbation obtained, I caused the clay model to be built of the natural size by measurement from the sketch model, and when it approximated to the form, I with my own hand in all instances secured the anatomical details and the characteristics of its nature.[1]

The next step was to make a copy in clay of the proof model, of the natural size of the extinct animal: the largest known fossil bone, or part, of such animal being taken as the standard according to which the proportions of the rest of the body were calculated agreeably with those of the best preserved and most perfect skeleton. The model of the full size of the extinct animal having been thus prepared and corrected by renewed comparisons with the original fossil remains, a mould of it was prepared, and a cast taken from this mould, in the material of which the restorations, now exposed to view, are composed.

As a backdrop to Hawkins' sculptures, the appropriate rock was brought in to indicate the type of strata in which the relevant fossils were found, and thus the period in which the creatures in question lived.

In respect of the two *Iguanodons*, materials used for each one included:

4 iron columns 9 foot long by 7 inches diameter; 600 bricks; 650 5-inch half-round drain tiles; 900 plain tiles; 38 casks of cement; 90 casks of broken stone; making a total of 640 bushels of artificial stone. These, together with 100 feet of iron hooping and 20 feet of cube inch bar, constitute the bones, sinews, and muscles of this large model.[2]

(Bushel: a measure of capacity equal to 8 gallons. Bar: long rigid piece of metal used as a fastening.[3])

Owen noted, however, that in cases where 'only the fossil skull and a few detached bones of the skeleton have been discovered', it was left to the discretion of Hawkins to assume 'the responsibility of adding the trunk', a process that would, of course, be to a large extent conjectural.[4]

The queen took a keen interest in the proceedings, and on 18 November 1853, she and her consort, Prince Albert, visited Hawkins' workshop, 'where the Royal visitors were surprised at the antediluvian wonders ... the vestiges and tracing of an earlier world'.[5]

The Times newspaper reported that a bank was currently being created in the park, on which 200,000 people spectators 'can sit and have a perfect view of the display of fountains more than 4 times that of Versailles [former principal royal residence of France]'.[6]

On 10 June 1854, Her Majesty returned to perform the opening ceremony of the new Crystal Palace and its grounds, in the presence of some 40,000 spectators.[7]

John Noble Wilford (born 1933), author and scientific journalist for the *New York Times*, stated that 'of all the prehuman animals, the creatures Owen called dinosaurs and Hawkins revived in stone became irresistible to the public and to an increasing number of scientists'.[8]

It was at the suggestion of the dean of Hereford that small-scale models of the sculptures be made available for educational purposes in schools across the land. Also, six large posters were created entitled 'Struggles of Life among the British Animals in Primeval Times'. 'Sheet three featured the famous *Iguanodons* in their Crystal Palace stances, and a *Hylaeosaur*.'[9]

In the summer of 1855, the directors of the Crystal Palace Company decided that, owing to lack of funds, Hawkins' work should be terminated. This was despite the fact that, as auctioneer and antiquary, Samuel Lee Sotheby pointed out:

> [Hawkins had] commenced his labours in his shed at the lower portion of the park. In the depth of winter, and in severest weather, that gentleman was always to be found at his post, and though receiving little more than journeyman's wages, he was content to work on and to carry out those wonderful reproductions of bygone ages, aided by the council of his friend, Professor Owen.[10] [Journeyman: a skilled worker who is employed by another.[11]]

Hawkins' sculptures for the Crystal Palace included four dinosaurs:

- *Iguanodons* (two): Suborder Ornithopoda, herbivore, quadruped (Early Cretaceous Period); beside the *Iguanodon* sculptures, Hawkins placed cycads and zamia (palm-like plants of the genus Cycad) cast in concrete; this being considered their likely source of food.[12]
- *Hylaeosaurus* (one): Suborder Ankylosauria, herbivore, quadruped (Early Cretaceous).

- *Megalosaurus* (one): Suborder Theropoda, carnivore, biped (Mid-Jurassic).

This above *Megalosaurus* has the distinction of being mentioned in the opening passage of a novel by the writer Charles Dickens (1812–70), namely *Bleak House*, published in 1853: 'Implacable November weather. As much mud in the streets as if the waters had but newly retired from the face of the earth, and it would not be wonderful to meet a *Megalosaurus*, 40 feet long or so, waddling like an elephantine lizard up Holborn Hill.' (Note: Hawkins may be forgiven for depicting his *Megalosaurus* as a quadruped; in fact, these creatures were bipedal.)

Hawkins' sculptures also included the following creatures, in order of their first appearance:

- *Labyrinthodon*: Amphibian (Late Palaeozoic to Early Mesozoic).
- *Dicynodon*: Therapsid, tortoise-like (Late Permian).
- *Ichthyosaurus*: Marine reptile, dolphin-like (Late Triassic to Early Jurassic).
- *Plesiosaurus*: Marine reptile, with elongated snake-like head and neck (Early Jurassic).
- *Teleosaurus*: Crocodilian, with an elongated snout (Mid-Jurassic).
- *Pterodactyl*: Winged reptile. Hawkins' two sculptured *Pterodactyls* were depicted sitting atop a rock, one with its enormous wings outstretched (Late Jurassic).
- *Mosasaurus*: Reptile, a true lizard, aquatic (Late Cretaceous).
- *Palaeotherium*: Mammal, ungulate (hoofed), tapir-like (Early to Mid-Eocene).
- *Anoplotherium*: Mammal, ungulate, similar to a camel but without its hump (Late Eocene).
- *Megatherium*: Mammal, the giant ground sloth (Early Pliocene to Early Holocene).
- *Megaloceros*: Mammal, a large early deer (Early to Late Pleistocene).

On 13 March 1868, Hawkins arrived in New York at the invitation of the commissioners of that city's Central Park. On 18 May, he was formally engaged by the commissioners 'to reproduce the original forms of life inhabiting the great continent of America'.[13] This would include the ornithopod dinosaur *Hadrosaurus*, and the tyrannosauroid dinosaur *Laelaps* (known today as *Dryptosaurus*).[14]

Sadly, Hawkins' studio was broken into and vandalised. Nonetheless, he was able to cast three *Hadrosaurus* skeletons: one for Princeton University, another for the Smithsonian Institution, and a third for the 1876 Philadelphia Exhibition. He also produced seventeen large paintings of prehistoric scenes for Princeton.

In 1936, the Crystal Palace burnt down. In 1986, the ownership of Crystal Palace Park was transferred to the London Borough of Bromley, where Hawkins' famous sculptures are now protected by law.

Conan Doyle and *The Lost World*

In 1912, when author Sir Arthur Conan Doyle (1859–1930) was aged 52, his novel *The Lost World* was published in the USA and in the UK. It was serialised in the *Strand Magazine* and published by Hodder & Stoughton. In that book was mention of the word 'dinosaur', and in writing it, the author demonstrated that he had more than a passing knowledge of those extinct creatures.

Conan Doyle was born in Edinburgh, Scotland, on 22 May 1859. Of Irish Catholic descent, he himself became a spiritualist. He studied medicine at Edinburgh University's Medical School and botany at the city's Royal Botanic Gardens. Having worked as a ship's doctor aboard a whaling vessel, he became a medical practitioner in Plymouth, Portsmouth, and, finally, London. His first novel featuring Sherlock Holmes, *A Study in Scarlet*, was published in 1887.

The following is an extract from *The Lost World*:

> Edward Malone, a young journalist for the *Daily Gazette*, is told by Gladys, with whom he has fallen in love, that her ideal man would be someone who 'could look Death in the face and have no fear of him, a man of great deeds and strange experiences'. Accordingly, Malone persuades Mr McArdle, his news editor, to send him 'on some mission for the paper'. This, it transpires, is an expedition to the Amazon region of South America, in company with Professor Challenger, the famous zoologist.
>
> Meanwhile, Malone refers to a difference of opinion that existed between Darwin and German evolutionary biologist August F.L. Weismann (1834–1914), who was, in fact, a strong supporter of Darwin's theory of evolution.
>
> Challenger had visited the Amazon region before, and had come across the sketch-book of a recently deceased explorer from the USA

named Maple White, which contained 'a double page of studies of long-snouted and very unpleasant saurians'. [Saurian: any large reptile, especially a dinosaur or other extinct form.[15]]

On his return home, Challenger referred to 'an excellent monograph by my gifted friend Ray Lankester' that contained an illustration, beneath which was written the inscription 'Probable appearance in life of the Jurassic dinosaur *Stegosaurus*. The hind leg alone is twice as tall as a full-grown man'.

Here, Conan Doyle demonstrates that he is not averse to mixing fact with fiction, British biologist Sir Edwin Ray Lankester (1847–1929) being a contemporary of his.

Challenger showed Malone a bone, also one of the deceased Maple White's possessions, which he declared, was 'no fossil specimen, but recent. It belongs to a very large, a very strong, and, by all analogy, a very fierce animal which exists upon the face of the earth, but has not yet come under the notice of science'. He also showed Malone a photograph that he had taken in Amazonia, 'of the *Dimorphodon*, or *Pterodactyl*, a flying reptile of the Jurassic Period', and stated 'that both the *Pterodactyl* and the *Stegosaurus* are Jurassic'.

When Challenger addressed London's Zoological Institute and opined that 'Creatures which were supposed to be Jurassic, monsters who would hunt down and devour our largest and fiercest mammals, still exist', he was treated with derision.

It was decided that Mr Summerlee, 'the veteran Professor of Comparative Anatomy', and Lord John Roxton, a 'sportsman and a traveller', who like Challenger had also previously visited Amazonia, should join the expedition. On its arrival in South America, two local 'swarthy fellows', Gomez and Manuel; a 'devoted negro' called Zambo; and some local Indians also joined the expedition.

In his novel, Conan Doyle voiced his admiration for men of science, and the difficulties they face, vicariously through the words of Professor Challenger, who stated as follows:

Every great discoverer has been met with the same incredulity – the sure brand of a generation of fools. When great facts are laid before you, you have not the intuition, the imagination which would help you to understand them. You can only throw mud at the men who have risked their lives to open new fields to science. You persecute the prophets! Galileo! Darwin, and I – .

This statement was followed by 'prolonged cheering'. Doubtless, Malone also echoed Conan Doyle's thoughts when he declared 'that both Summerlee and Challenger possessed that highest type of bravery: the bravery of the scientific mind. Theirs was the spirit which upheld Darwin among the gauchos of the Argentine, or Wallace among the head-hunters of Malaya'.

On venturing into the interior, it was not long before the explorers caught sight of 'a huge gray bird', which Challenger subsequently shot. He described it as a *Pterodactyl*, though Summerlee remained to be convinced. That was, until in the darkness, as the party were cooking an ajouti, 'a small, pig-like animal', another *Pterodactyl* arrived on the scene, and snatched their meal. Said Malone: '... suddenly out of the darkness, out of the night, there swooped something with a swish like an aeroplane. The whole group of us were covered for an instant by a canopy of leathery wings, and I had a momentary vision of a long, snake-like neck, a fierce, red, greedy eye, and a great snapping beak, filled, to my amazement, with little, gleaming teeth. The next instant it was gone – and so was our dinner. A huge black shadow, twenty feet across, skimmed up into the air; for an instant the monster wings blotted out the stars, and then it vanished over the brow of the cliff above us'.

When Roxton noticed 'an enormous three-toed track' which was 'imprinted in the soft mud', the party believed, at first, that it had been made by an enormous bird. 'Smaller tracks of the same general form were running parallel to the large ones. 'But what do you make of this?' cried Professor Summerlee, triumphantly, pointing to what looked like the huge print of a five-fingered human hand appearing among the three-toed marks. 'Wealden!' cried Challenger, in an ecstasy. 'I've seen them in the Wealden clay. It is a creature walking erect upon three-toed feet, and occasionally putting one of its five-fingered forepaws upon the ground. Not a bird, my dear Roxton – not a bird.'

'A beast?'

'No; a reptile – a dinosaur. Nothing else could have left such a track. They puzzled a worthy Sussex doctor some ninety years ago; but who in the world could have hoped – hoped – to have seen a sight like that?'

Here, Conan Doyle has in mind paleontologist Dr Gideon Mantell:

> Whereupon, said Malone, in 'an open glade ... were five of the most extraordinary creatures that I have ever seen. Crouching down among the bushes, we observed them at our leisure. There were ... five of them, two being adults and three young ones. In size they were enormous. Even the babies were as big as elephants, while the two large ones were far beyond all creatures I have ever seen. They had slate-colored skin, which was scaled like a lizard's and shimmered where the sun shone upon it. All five were sitting up, balancing themselves upon their broad, powerful tails and their huge three-toed hind feet, while with their small five-fingered front-feet they pulled down the branches upon which they browsed'. They looked like 'monstrous kangaroos, twenty feet in length, and with skins like black crocodiles', and Challenger identified them as *Iguanodons*.
>
> Further on, in an enormous pit with 'pools of green-scummed, stagnant water, fringed with bullrushes was a rookery of *Pterodactyls*. Large and small, not less than a thousand of these filthy creatures lay in the hollow before us', said Malone.
>
> When a 'monstrous organism' with 'two terrible, greenish eyes' approached the camp fire, Roxton drove it away with the aid of a 'blazing branch'. The following morning, they discovered that 'the *Iguanodon* glade was the scene of a horrible butchery. One of these unwieldy monsters ... had been literally torn to pieces by some creature not larger, perhaps, but far more ferocious, than itself'. Challenger suggested that the attacker may have been an *Allosaurus* dinosaur. Or a *Megalosaurus*, said Summerlee. Challenger agreed. 'Exactly. Any one of the larger carnivorous dinosaurs would meet the case. Among them are to be found all the most terrible types of animal life that have ever cursed the earth or blessed a museum'.
>
> Malone stated that his eyes 'chanced to light upon the enormous gnarled trunk of the gingko tree which cast its huge branches over us'.

Here, again, Conan Doyle displays his knowledge of botany, for fossils of the gingko date back 270 million years, to the Permian Period of the Late Palaeozoic Era:

> Malone had named a stretch of water 'Lake Gladys', after his lady-love. Here, he encountered 'a new-comer, a most monstrous animal ... coming down the path. For a moment I wondered where I could

have seen that ungainly shape, that arched back with triangular fringes along it, that strange bird-like head held close to the ground. Then it came back, to me. It was the *Stegosaurus* – the very creature which Maple White had preserved in his sketch-book, and which had been the first object which arrested the attention of Challenger! There he was – perhaps the very specimen which the American artist [Maple White] had encountered. The ground shook beneath his tremendous weight, and his gulpings of water resounded through the still night. For five minutes he was so close to my rock that by stretching out my hand I could have touched the hideous waving hackles upon his back. Then he lumbered away and was lost among the boulders'.

Malone subsequently 'stood like a man paralyzed', when 'there was movement among the bushes at the far end of the clearing which I had just traversed. A great dark shadow disengaged itself and hopped out into the clear moonlight. I say 'hopped' advisedly, for the beast moved like a kangaroo, springing along in an erect position upon its powerful hind legs, while its front ones were held bent in front of it. It was of enormous size and power, like an erect elephant, but its movements, in spite of its bulk, were exceedingly alert. For a moment, as I saw its shape, I hoped that it was an *Iguanodon*, which I knew to be harmless, but, ignorant as I was, I soon saw that this was a very different creature'.

Conan Doyle was clearly well aware that *Iguanodons* were herbivorous dinosaurs:

Instead of the gentle, deer-shaped head of the great three-toed leaf-eater, this beast had a broad, squat, toad-like face like that which had alarmed us in our camp. His ferocious cry and the horrible energy of his pursuit both assured me that this was surely one of the great flesh-eating dinosaurs, the most terrible beasts which have ever walked this earth. As the huge brute loped along it dropped forward upon its forepaws and brought its nose to the ground every twenty yards or so. It was smelling out my trail.

After further harrowing adventures, the party returned to England, where Professors Summerlee and Challenger both addressed members of the Zoological Institute; only to be greeted with immense scepticism. That was until, 'a large square packing-case' was brought onto the platform containing a real, live *Pterodactyl*!

Malone's story did not end happily. Gladys scolded him for abandoning her, and announced that in his absence, she had married a solicitor's clerk![16]

The Lost World shed light on the creative and knowledgeable character of Conan Doyle, whose 1912 novel was to inspire countless other books and films of the science fiction genre.

Chapter 7

Mary Anning: Fossil Hunter Extraordinaire!

Mary Anning was born on 21 May 1799 at Lyme Regis ('Lyme'), a coastal town in west Dorset. Her father, Richard, was a cabinet-maker/carpenter and married to Mary 'Molly' (*née* Moore) of Blandford, Dorset.

The Annings' house was on the seafront at Lyme, on the site of the present town's museum. The museum was commissioned in 1901 by Thomas Philpot, who was a relative of fossil collector Elizabeth Philpot (1780–1857).

Richard Anning supplemented his income by collecting and selling fossils. His daughter, Mary, and her elder brother, Joseph, accompanied him on his expeditions. Richard died in 1810 when Mary was aged 11, after which time her education, 'so far as is known', came to an end.[1]

Lyme's fossils are of marine origin. They were laid down on the seabed between 200 and 195 mya (Early Jurassic), when the land lay at a latitude of about 30 degrees north, in a correspondingly warm climate. The fossils are mainly to be found in a geological formation known as the Blue Lias (a mixture of shale and limestone).

A companion of Mary's, during her fossil-hunting excursions, was Henry de la Beche, a member of the local gentry. In 1835, de la Beche founded the Geological Society of Great Britain. In 1842, he would be knighted for services to the science of geology.[2]

Mary's 'list of significant discoveries' spanned the period 'between the *Ichthyosaur* which her brother Joseph and she put together in 1811 to 1812, to her last major find of a *Plesiosaur* – *Plesiosaurus microcephalus* – in 1830'. The list included 'at least 3 completed Ichthyosaurs, 2 Plesiosaurs [and] a quantity of coprolite ... which she correctly identified as fossilized faeces'.[3]

Other distinguished visitors who were drawn to Lyme on account of its fossils included the following:

William Buckland – The aforementioned was the first to write (and publish) a full and scientific description of a dinosaur. This was a megalosaur, the fossilised remains of which he had discovered near Oxford in 1824 (as already mentioned, Richard Owen did not coin the term 'dinosaur' until 1841). The megalosaur now bears his name, *Megalosaurus bucklandi*.

William Conybeare – Anglican clergyman, geologist and palaeontologist, who was 'best known for his ground-breaking work on marine reptile fossils in the 1820s'.[4]

Louis Agassiz – Swiss palaeontologist.

Sir Philip Egerton – Conservative politician and palaeontologist.

Lord Enniskillen – Irish Conservative politician and palaeontologist.

The King of Saxony[5] – This was Frederick Augustus II, who reigned from 1836 to 1854. In 1844, accompanied by his personal physician, he made an informal visit to England and Scotland. Among the places he visited was Lyme, 'where he purchased from the local fossil collector and dealer, Mary Anning, an *Ichthyosaur* skeleton for his own extensive natural history collection. It was not a state visit, but the King was the guest of Queen Victoria and Prince Albert at Windsor Castle'.[6]

Thomas Birch – A former soldier and wealthy fossil hunter from Lincolnshire, who purchased many fossils from the Anning family. In 1820, he sold his collection for £400 and donated the proceeds to:

> … the poor woman [Molly Anning] and her son [Joseph] and daughter [Mary] at Lyme who have in truth found almost *all* the fine things [fossils], which have been submitted to scientific investigation: when I went to Charmouth & Lyme last summer [1819] I found these people in considerable difficulty – in the act of selling their furniture to pay their rent – in consequence of their not having found one good fossil for nearly a twelve month'.[7]

In that same year of 1820, Mary turned part of the family home into a shop, where she offered for sale fossils that she had found. Gideon Mantell, after his visit to Lyme in June 1824, described it disparagingly as 'a little dirty shop with hundreds of specimens piled around in the greatest disorder'.[8] Remembering the 1840s, a resident of Lyme was more charitable, and described ammonites, which Mary 'washed and burnished till they shone like metal'.[9]

Although Mary's fossils found their way into various museums and collections, she was seldom given the credit as their discoverer, nor for her extensive knowledge of the subject. Anna Pinney, who accompanied Mary on her fossil-hunting expeditions, wrote: 'She says the world has used her ill. These men of learning have sucked her brains and made a great deal of publishing [their] works, of which she furnished the contents, while she derived none of the advantages.'[10]

However, in her later years, the British Association for the Advancement of Science in Dublin, and the Geological Society of London, raised money for Mary by private subscription so that she should not be 'in want'.[11]

Mary never married. She died on 9 March 1847 and was buried in the churchyard of the nearby Parish Church of St Michael.

Mary Anning was a truly remarkable person. Largely self-educated, intelligent, industrious and enterprising, she demonstrated that even 'a mere woman', and an amateur at that, could earn the respect of the foremost palaeontologists in the land. Although her discoveries were mainly of marine reptiles and not dinosaurs, she fired the imagination of the public, and showed that people of all classes could become fossil hunters.

Finally, the fate of the marine reptiles was largely bound up with the fate of the dinosaurs, as will shortly be seen.

Chapter 8

Dinosaur Prints: How to Identify Them

Ichnotaxonomy; Prints and Casts

Ichnotaxonomy is a system of scientific classification used to identify prints, tracks and traces of ancient animals.

The Quality of Prints and Casts

Fossil hunters may be fortunate enough to find distinct and well-preserved track impressions, but whether a dinosaur's print is a faithful representation of what it originally looked like depends on a number of factors:

- The substrate (surface on which the creature walks or runs): the texture and homogeneity of the substrate and its moisture content and firmness is of crucial importance. It is obvious, for example, that a print made in not overly moist clay is far more likely to retain its shape than one made in dry or moist sand. The best-preserved prints, said Tony Thulborn, zoologist of Queensland University, Australia, occur in fine-grained sediments such as siltstones, mudstones and sandstones.[1]
- Subsequent damage or distortion by upheavals in the Earth's crust.
- Subsequent weathering of the imprints.
- Bioturbation (the disturbance of sedimentary deposits by living organisms[2]).
- Transmitted prints (*aka* 'ghost', 'under', or 'over' prints): It may be difficult to interpret 'transmitted prints' – i.e. where the impact of the hind foot or forefoot was 'transmitted through a succession of sediment layers to form a stack of casts and moulds'.[3]
- Eroded or displaced prints: The forefoot prints of quadrupeds are smaller and usually shallower than the hind-foot prints, because less weight is placed on them. They are therefore more vulnerable to erosion. Also, prints may be distorted or displaced by subsequent movement of the Earth's crust.
- Overlapped prints: In the case of a quadruped, as it walks, its rear feet may obliterate the impressions made by its forefeet. This may

lead to the mistaken conclusion that the creature is a biped. Confusion may also arise if several dinosaurs of the same or different species leave their prints in the same terrain. Alternatively, the fossil hunter may discover the 'natural cast': formed by the infilling of the individual prints with sediment. This will therefore be convex in shape. Often both print and cast are found at the same time, when the rock containing them is split apart.

How Dinosaur Prints are Formed

Clearly, when a print is formed, it is not the bones of the dinosaur's forefoot or hind foot that come into contact with the underlying substrate but the skin, beneath which lie fleshy pads. Thulborn said:

> In some cases the digits are U-shaped with parallel sides; and in others they are angular or V-shaped with pointed tips. Occasionally, the digits comprise a series of swellings or nodes. These nodes mark the presence of fleshy pads or cushions under the toes. In deeply impressed footprints, the digital nodes may be contiguous, separated only by weak constrictions. But in shallow footprints, where the dinosaur's foot did not sink so deeply into the substrate, each digit may be represented by a series of discrete nodes. At the tips of the digits there are often indications of claws, which may be long or short, narrow or broad, straight or curved, V-shaped or U-shaped.[4]

US dinosaur-track worker and enthusiast Glen Kuban explained how a well-preserved and distinct dinosaur print or track may be formed:

> First, a trackmaker walks along a moist but firm, fine-grained sediment. In most cases, the tracks remain exposed for a short while, allowing them to become drier and harder (and thus able to resist damage during subsequent burial). A short time later the prints are gently buried with additional sediment, preferably of a contrasting type (which would allow the layers to separate when later re-exposed). While buried for millions of years, the original sediment lithifies (turns into rock).[5]

Clearly, the stance adopted by the dinosaur, whether it is semi-erect or erect, will affect the shape of its prints; as will its posture, according to whether it is walking, trotting or running, which will also determine the length of its stride (distance travelled by the same foot between consecutive footsteps). The stance and posture, therefore, determine the manner in which the forefeet and hind feet interact with the substrate, and also determine the width of the trackway.

Techniques used to Analyse Dinosaur Prints

They include photography, measurements using grids, mould and cast making, Lidar, the use of drones, and 3D printing. (Lidar: a detection system that works on the principles of radar but uses light from a laser.[6])

Was the Trackmaker Bipedal or Quadrupedal?

Early Saurischians were bipedal; some such as theropods later became bipedal, while others such as sauropods became quadrupeds. Early Ornithischians were bipedal, too, but many subgroups, including stegosaurs, ankylosaurs and ceratopsians, became quadrupedal, as did some ornithopods, while other ornithopods remained largely bipedal, as some Cretaceous trackways show.[7]

However, Thulborn issued a caveat in regard to generalisations, saying: 'It should be remembered that not all genuses of dinosaur were exclusively either bipedal or quadrupedal'.[8]

Shape of Prints as a Clue to the Identity of the Trackmaker

The prints of the following subgroups included the following features:

Sauropods
- Shape of forefoot print: Roundish.
- Number of fingers: Five.
- Shape of fingers: Blunt and spread widely around the print.
- Size of hind-foot print: Larger than the forefoot print; some hind-foot prints are more than a metre long.
- Shape of hind-foot print: Large and oblong-shaped, resembling large bear prints.
- Number of toes: Five, of decreasing size from the inside to the outside of the foot.
- Shape of toes: Kuban states: 'Generally, the first (inner) three digits are prominent and bear large, sharp claws, which often record well in tracks, with the fourth and fifth digits usually smaller, blunter, and clawless'.[9]
- Presence of claws: Only the inner three or four digits (depending on the species) bore large claws, which record well in tracks. The

fourth and fifth digits were generally small and clawless, but often detectable in tracks.

Note: Often the front prints were overlapped by the rear prints, or mud was pushed forward by them, either obliterating them or reducing them to crescent-shaped depressions.

Theropods
- Shape of hind-foot print: Print is longer than it is wide; posterior end is somewhat V-shaped.
- Number of toes: Three large toes and one smaller, rear-facing digit (called the hallux) that rarely registers, since it is normally held fairly high on the foot.[10] Therefore, theropod tracks are described as 'tridactyl'. As Kuban's research has shown, 'in some theropod tracks, especially ornithomimids [group with a superficial resemblance to modern-day ostriches], the hallux and all, or part of the metatarsus (sole and heel) records, apparently because they sometimes walk in a plantigrade (heel-down) fashion, perhaps reflecting a crouching posture, while foraging for small food items in mud or shallow water'.[11]
- Shape of toes: Long, narrow and pointed. Kuban states: '... the pad pattern on well-preserved theropod tracks is often quite diagnostic. That is, they show two, three, and four pads on digits two, three, and four respectively. Also, the hind-foot prints often show a small indentation or "instep" on the inside of the foot, which along with claw features, allows one to know if a print is a right or a left, even if not in a striding trail.'[12]
- Presence of claws: Digit impressions terminate with sharp, slender claw marks. Kuban states: '... the claws in digits two and three (the inside-most and central digit) usually point towards the inside of the foot, which like the instep, helps one identify a left from a right, even when isolated'.[13]

Kuban points out that 'tracks in ornithopod trails are typically pointed inward or "pigeon-toed", in contrast to theropod tracks which usually point straight ahead'.[14]

Ornithopods
- Shape of forefoot print: Irregular and 'often resemble a smaller version of the hind-foot prints, sometimes with one or two additional digits recorded'. However, forefoot prints 'may or may not be

recorded, since many ornithopods alternate between quadrupedal and bipedal gaits, and the forefoot prints are generally impressed more shallowly than the hind-foot prints; if recorded at all'.[15]
- Number of fingers: Five, but often only the three largest ones register.
- Shape of fingers: Typically short, blunt.
- Size of hind-foot print: Larger than forefoot print.
- Shape of hind-foot print: 'Generally wider than long, with well-rounded posteriors, and a more symmetrical overall shape than theropod prints (with no instep)'. Also, 'ornithopod hind[-foot] prints sometimes show a small, flattened area or inward "notch" at the back'.[16]
- Number of toes: Three large toes and one smaller, rear-facing digit that rarely registers in a track. Therefore, like theropod prints, ornithopod prints are described as 'tridactyl'.
- Shape of toes: Wide, stubby and tapering.

Estimating the Size of a Dinosaur from the Size of its Prints

Thulborn said: the 'dimensions of the footprints can be used to estimate its body size'.[17] But this was problematical, and he concluded that the best method, where possible, was to compare the print with the known shape of the forefeet and hind feet of various fossilised dinosaur skeletons, whereupon a three-dimensional, computer-generated model of that particular dinosaur may be created. Finally, 'on the assumption that the specific gravity of the dinosaurian body was roughly equal to that of water, it is possible to estimate the weight of the dinosaur'.[18]

It is now time to study the various dinosaur prints that have been discovered in the Isle of Purbeck over the years.

Chapter 9

Dinosaur Prints: Some Local Discoveries

A number of dinosaur tracks [i.e. prints] and trackways have been identified in the Purbeck (Limestone) Group. The majority of these tracks were made by three-toed [tridactyl: having three toes or fingers.[1]] bipeds, and they are frequently attributed to '*Iguanodon*-like' or '*Megalosaurus*-like' trackmakers. However, variations in preservation and sedimentary dynamics can make identification of trackmakers difficult and in many cases, it may not be possible to distinguish ornithopod and theropod tracks unless preservation allows recognition of taxon-specific features (e.g. the characteristic fore [foot] prints of iguanodontian ornithopods or the narrow claw marks expected in a theropod).[2]

Keates' Quarry, Acton (1997)

The Keates family has been working stone in Purbeck since 1698 and at Keates' Quarry, Acton, since 1951. It was here, in January 1997 – on a sloping pavement or 'horizon' (a layer of rock with particular characteristics or representing a particular period[3]) of Purbeck stone, about half the size of a football pitch – that the remarkable discovery of no less than 111 dinosaur prints was made.

The quarry, which lies some 400 feet or so above sea level, is 2½ miles to the west of Swanage and south of the road leading from Langton Matravers to Worth Matravers. Quarry proprietor Kevin Keates was excavating layers of Purbeck stone when Treleven Haysom, proprietor of nearby St Aldhelm's Quarry, noticed indentations in the rock surface that had been exposed. These, he identified as dinosaur prints.

In a paper published in the *Proceedings of the Dorset Natural History & Archaeolgical Society* in 1998, entitled 'Keates' Quarry Dinosaur Footprint Site: Intermarine Member, Purbeck Limestone Group (Berriasian)' (in cooperation with the local landowner, the National Trust), Dr Joanna Wright of the Department of Geology, Bristol University, reported:

> The fossil footprints are preserved in a typical Purbeck shelly [highly fossiliferous] limestone deposited on the edge of a freshwater lagoon

which was separated from the open sea, by some kind of barrier. At the time of the track formation this area was probably a beach. The bed overlying the footprint surface was deposited rapidly, possibly by a storm.

There are 111 footprints on the exposed trackway surface. The tracks were made by more than one individual – probably more than twelve. Most of the footprints are roughly oval in shape. Thirteen footprints on the exposed surface are D-shaped. The footprints vary considerably in depth. No claw marks or pad impressions are evident on any of these tracks.

The combination of sedimentological evidence from the raised rims and from the footprint infill indicates that these footprints are primary prints. That is to say that the limestone surface now preserved is the actual surface on which the dinosaurs walked. It is likely that the dinosaurs walked over the sediment surface when it was exposed above the water level.

When the site was inundated again shortly afterwards, sediment was deposited 'very rapidly ensuring preservation of the footprints'.

Dr Wright declared that 'the very large size of some of these footprints' indicated that the trackmakers were sauropod dinosaurs – brachiosaurids being the 'most likely' candidates.[4]

- Identity: genus *Brachiosaurus* (order Saurischia; suborder Sauropodomorpha; clade Sauropoda; family Brachiosauridae).
- Period: Late Jurassic.

Other discoveries of dinosaur prints and tracks made in Purbeck include the following:

Suttle's Quarry, Herston Swanage (1933)

Fourteen prints were discovered by local archaeologist J. Bernard Calkin.[5]

- Identity: genus *Iguanodon* (order Ornithischia; suborder Ornithopoda; family Iguanodontidae).
- Period: Early Cretaceous.

Suttle's Quarry, Herston (1961)

In 1965, dinosaur enthusiast and collector Ernest F. Oppé wrote that in the summer of 1961, 'a marvel was contributed by Messrs. J. [John] and E.W. Suttle at their quarry; an old one re-opened on the height above Herston Cross, Swanage'. Oppé continued:

Mr John Suttle retained for inspection and duly reported to scientific centres, a display of two uninterrupted parallel lines of regularly spaced and pointing footprints; all of a well-known local character [i.e. similar to other discoveries in the neighbourhood], namely 3-digit impressions, and all of the same 'medium Purbeck' size – i.e. measurable within the circumference of a circle about 10 inches to 12 inches in diameter.

The two parallel lines were said to be '2 feet apart' and running 'south-west for 25–26 feet into the rock face of much unquarried material'.

In May 1962, another track resembling each of the two lines of the 'double track' was discovered running alongside it and 'about 2 full yards north of it'. In June 1963, the trackway was removed to London's Natural History Museum and put on display.[6] Calkin concluded that the tracks were made by a 'carnivorous dinosaur: a species of *Megalosaurus*'.[7]

- Identity: genus *Megalosaurus* (order Saurischia; suborder Theropoda; family Megalosauridae; subfamily Megalosaurinae.)
- Period: Mid-Jurassic.

Lock's Quarry, Acton, Langton Matravers (June 1967)

Two trackways were discovered, 'average size of the imprints in both trackways is 11 inches high and 13½ inches respectively'.[8] Was this 'Mr and Mrs Dinosaur' taking a walk together?

- Identity: genus *Megalosaurus*.

Queensground Quarry, Langton Matravers (1976)

Three trackways were discovered by proprietor Ronald Lewis. 'One trackway consists of footprints in which the antero-posterior and the lateral dimensions average 9 inches and 10 inches respectively. The two remaining trackways have imprints of more or less identical proportions, the equivalent dimensions being 7½ inches and 8½ inches respectively.'

Furthermore, said Justin B. Delair, author of *The Mesozoic Reptiles of Dorset* in three parts (1958–1960): 'Scattered bones of *Iguanodon* dinosaurs have been found in the Purbeck strata of Dorset on several occasions.'[9]

- Identity: genus '*Iguanodon*' type'.

Townsend Road, Swanage (1981)

Isolated prints were spotted in excavations for a new house. The site, which measured about 140 square yards, was excavated by a team from the

Dorset County Museum.[10] In excess of seventeen trackways, consisting of more than 170 tridactyl prints on four horizons, were exposed.[11]
- Identity: genus possibly *Iguanodon* and *Megalosaurus*.

Sunnydown Farm Quarry, Langton Matravers (1988)

Dinosaur tracks were discovered by geologist Paul C. Ensom, the Assistant Curator of Dorset County Museum.
- Identity: Sauropod and *Iguanodon*.[12] (Sauropod: order Saurischia; suborder Sauropodomorpha; clade Sauropoda.)

Durlston Bay, Swanage (1990)

A single track was discovered by anaesthetist and Egyptologist Dr John F. Nunn.
- Identity: unidentified tridactyl.[13]

Conclusion

The dinosaurs that roamed the area that is now known as Purbeck therefore included both herbivores and carnivores. As is always the case in nature, and probably always was, the strong prey upon the weak, and it is likely that the carnivorous theropods would have preyed on the herbivorous ornithopods and sauropods, as well as upon the small mammals and other reptiles. This possibility is suggested by the fact that bite marks have been discovered on the fossilised skeletons of the following dinosaurs: *Tenontosaurus*, *Triceratops* and *Hypacrosaurus* – all of which were herbivores.

To kneel on the stone surface, with the kind permission of quarry proprietor Kevin Keates, and place his hand on the indentation of a print that was made 140 or so million years ago by creatures that flourished on Earth for approximately 181 million years (247–66 mya), was, for the author, a truly mind-blowing experience. Furthermore, the very size that some of these creatures could attain is truly staggering. For example, in respect of the following, the approximate dimensions for a typical adult are as follows:

- *Megalosaurus*: theropod, carnivorous, length 30 feet, weight 1 ton.
- *Brachiosaurus*: sauropod, herbivorous, length 100 feet, weight 60 tons.
- *Iguanodon*: ornithopod, herbivorous, length 32 feet, weight 3 tons.

Dinosaur Prints: Some Local Discoveries

How does this compare with the largest terrestrial creatures alive today? The largest living reptile is the saltwater crocodile, with a length of 18 feet and a weight just under 1 ton. The largest mammal is the African bull elephant, with a length of 25 feet and a weight of 6 tons. So, the *Brachiosaurus* was sixty times heavier than the crocodile and ten times heavier than the elephant. The question now is, have any fossils of dinosaurs been discovered in Purbeck? The answer is yes.

Chapter 10

Some Dinosaur Fossils from Purbeck and Further Afield

Swanage Museum and Heritage Centre

A visit to this museum is rewarding because here are to be found various fossilised dinosaur bones that were discovered locally. For example, a vertebral body from the tail of a dinosaur, identified as an *Iguanodon*, which was discovered in the Wealden Beds, as already mentioned. Beside it, in the display cabinet, are several tail vertebrae of an unknown dinosaur, discovered not half a mile away at Peveril Point.

A member of staff kindly opened the cabinet and permitted the author to examine the vertebral body of the *Iguanodon*. It measured 3½ inches across in both directions and fitted into the palm of his hand. What a thrill it was for him after having studied these amazing creatures for so long. It was surprisingly heavy and abrasive to the touch.

The *Iguanodon* is a herbivorous, quadrupedal, ornithopod dinosaur, only two species of which are well documented: *I. bernissartensis* and *I. galvensis*. The former is estimated to have weighed a colossal 3.4 tons and measured up to 43 feet in length when fully grown.

- Identity: genus *Iguanodon* (order Ornithischia; suborder Ornithopoda; family Iguanodontidae).
- Period: Early Cretaceous.

The 'Square and Compass'

This eighteenth-century, stone-roofed public house is 3½ miles west of Swanage at Worth Matravers. Inside is a museum – which was created by its late publican Charlie Newman (1871–1953) – containing many fossilised relics of the past; for example, there are ten *Iguanodon* vertebrae, including five from the tail.

The Etches Collection

Some 7 miles to the west of Swanage, in the hamlet of Kimmeridge, a mile or so from the sea, is the Etches Collection of fossils, about which more will be said shortly. The fossils were collected locally; the local strata being Kimmeridge Clay, which is a marine sedimentary deposit and, therefore, not one in which dinosaur fossils would normally be found. However, some did find their way to the bottom of the ancient sea, presumably having been washed into it by coastal flooding or tsunamis. Finds include the humerus, fibula, femur and tail chevrons (V-shaped bony projections pointing downwards) of a sauropod and the claw of an unknown theropod dinosaur.

The Etches Collection also contains a vertebra and dermal scute (thickened bony plate) of a dinosaur of the genus *Stegosaurus*. The Stegosauria, however, lived during the period Mid-Jurassic to Early Cretaceous and had, therefore, become extinct well before the KT extinction event.

Stegosaurus belongs to the group Thyreophora, the members of which are characterised by protective 'armour' plates, or 'osteoderms' – bony deposits forming scales, plates or other structures, based in the skin (dermis).

Dinosaurs of the genus *Stegosaurus* were an estimated 30 feet in length and weighed up to 3 tons. These creatures were remarkable in that a double line of protective bony plates projected vertically from the neck, back and tail.
- Identity: genus *Stegosaurus* (order Ornithischia; group Thyreophora; suborder Ankylosauria; family Stegosauridae; subfamily Stegosaurinae).
- Period: Late Jurassic.

The Dorset County Museum

The museum, in Dorchester, Dorset's county town, is owned and managed by the Dorset Natural History & Archaeological Society. Here are to be found the following specimens, from various locations.

Durlston Bay, Swanage

In 1852, Swanage resident and amateur palaeontologist Charles Willcox discovered 'a small fragmentary left dentary' of a theropod dinosaur in the Cherty Beds of the Feather Quarry, Durlston Bay. (Dentary: bone of the lower jaw that bears the teeth.[1] Cherty Beds: now known as the Portland Chert Member. Chert: a hard, dark, opaque rock composed of silica with an amorphous or microscopically fine-grained texture.[2])

The chert was originally a sedimentary freshwater deposit, laid down in a lake, which was subsequently inundated by the sea. The dentary was about 1½ inches long and bore nine teeth. It was described and named '*Nuthetes destructor*' by Sir Richard Owen in 1854.

Robert Coram said this creature was 'a small flesh-eater with sharp, serrated teeth'. The nature of the teeth is an indication of dietary habits. A carnivore possesses sharp, incisor-like teeth for cutting, whereas a herbivore possesses rounded, molar-type teeth for chewing and grinding.

Coram continued: 'We can guess that *Nuthetes* was an agile creature between 1 and 2 metres long [3–6½ feet], with a fairly small head and long hind legs and tail. It probably prowled the margins of lakes scavenging for dead fish or scampering through undergrowth in search of lizards and ... tiny mammals.'[3]

- Identity: genus *Nuthetes* (order Saurischia; suborder Theropoda; family Dromaeosauridae).
- Period: Early Cretaceous.

Durlston Bay, Swanage

The fossilised remains of 'one of the smallest dinosaurs, well under a metre in length ... a plant-eater with tiny spiky teeth', was discovered in the Purbeck Beds of Durlston in 1857 by lawyer and fossil hunter Samuel Husbands Beckles.[4] In 1861, Richard Owen named it *Echinodon becklesii* after its discoverer. It is the only known species of this creature.

Echinodon weighed an estimated 5–10 pounds. It was bipedal and carnivorous. Unlike most ornithiscians, its teeth included one or two canines in each maxilla (left and right upper jaw).

- Identity: genus *Echinodon* (order Ornithischia; subgroup Echinodon ('hedgehog tooth'); family Heterodontosauridae ('different toothed lizards')).
- Period: Early Cretaceous.

Elsewhere in Dorset

The Cliffs of Black Ven near Charmouth, Dorset

In 1858, the almost complete remains of a fossilised dinosaur were discovered by quarry proprietor James Harrison of Charmouth, while he was quarrying at the above location. It was found in a marine limestone deposit, suggesting that this terrestrial creature had been washed out to sea. The specimen was described by Richard Owen in 1861. He named it *Scelidosaurus harrisonii*, after its discoverer.

Scelidosaurus, a quadrupedal dinosaur, weighed about a quarter ton and was 13 feet or so in length. Coram said this 'armoured dinosaur' typically

had 'a protective covering of bony plates and spines' along its back, head and tail.[5]
- Identity: genus *Scelidosaurus* (order Ornithischia; clade Thyreophora; family Ankylosauridae; Scelidosauridae).
- Period: Early Jurassic.

Smallmouth Sands, Weymouth, Dorset

The left humerus of a sauropod dinosaur was discovered in the Kimmeridge Clay Formation at this location by R. I. Smith in the early 1870s. In 1873, it was purchased by the Natural History Museum. It was subsequently identified as a basal titanosauriform and given the name *Durititan*.[6] (Basal: primitive, ancestral, the group that gave rise to later forms.)
- Identity: genus *Durititan* (order Saurischia; clade Sauropoda/ Titanosauriformes).
- Period: Late Jurassic.

Durlston Bay, Swanage

In 1874, Richard Owen 'described a small right dentary' from the 'Middle Purbeck Beds' at this location. This was a new species, which he named *Iguanodon hoggii*. In 2009, it was renamed *Owendon hoggii* in Owen's honour.[7]

Gillingham, Dorset

The partial right forelimb of a stegosaurian dinosaur was discovered in the Kimmeridge Clay Formation at Gillingham in 1938. It was presented to the Natural History Museum and given the name *Dacentrurus armatus*.
- Identity: genus *Dacentrurus* (order Ornithischia; suborder Stegosauria; family Stegosauridae).
- Period: Late Jurassic.

Kimmeridge, Dorset

In 1984, Peter Langham, fossil collector of Bridport, discovered a partial dinosaur skeleton (excluding skull and forelimbs) in the Upper (Late) Kimmeridge Clay Formation on the coast near Kimmeridge. Bipedal, feathered, and about the size of an ostrich, it was given the name *Juratyrant langhami*, after its discoverer.[8]

A model of *Juratyrant*, made by author, science writer and dinosaur expert Dougal Dixon, may be seen in Wareham Town Museum. Dixon said: '... and now we are pretty sure that there were more feathers than on my model. I gave it a naked face and naked legs to channel [emulate] a

vulture or an ostrich. Now we think that the feathers went all the way to the toes, and the face was covered too.'[9]
- Identity: Species *Juratyrant langhami*; genus *Juratyrant* (order Saurischia; suborder Theropoda; superfamily Tyrannosauridea).
- Period: Late Jurassic.

Dorset's Dinosaurs: Summary

Writing in 2011, Paul M. Barrett and Susannah C.R. Maidment, Department of Palaeontology, Natural History Museum, London, stated as follows, in respect of 'valid taxa', relating to the evidence for dinosaurs in Dorset.

Ornithischians

'Only four valid taxa are recognised from the county with certainty'. These are *Dacentrurus* (stegosaur); *Echinodon* (heterodontosaurid); *Owenodon* (iguanodontian ornithopod); and *Scelidosaurus* (basal thyreophoran).

However, 'More fragmentary material also indicates that ankylosaurs, small ornithopods, and other *Iguanodons* would also have been present.'

Saurischia

Sauropodomorpha (sauropods). Only one valid taxon is recognised, namely *Duriatitan* (titanosauriform).

Theropoda (theropods). Four valid taxa are recognised: *Magnosaurus* and *Duriavenator* (megalosaurids); *Metriacanthosaurus* (allsauroid); *Stokesosaurus* (tyrannosauroid); and *Nuthetes* (dromaeosaurid).[10] (*Stokesosaurus* is a reference to *Stokesosaurus langhami*, which has now been reclassified as *Juratyrant langhami*; see above).

Chapter 11

How is the Age of the Fossilised Bone of a Dinosaur Ascertained?

The ultimate source of food for all animals is plants. Plants take up carbon dioxide from the atmosphere, including radioactive carbon-14 (C-14). (Atmospheric C-14 is constantly being renewed by the bombardment of atmospheric nitrogen by cosmic rays.)

Once an animal dies, the C-14 that has become assimilated into its body begins to decay. But because of the rate of this decay, radiometric C-14 dating (radiometric: relating to the measurement of radioactivity) is only useful if the fossilised bone samples are less than 50,000 years old. C-14 is, therefore, of no value when it comes to determining the age of dinosaur fossils.

However, isotopes such as uranium-238, uranium-235 and potassium-40 have a far slower rate of decay. But these elements are not typically found in dinosaur fossils, which form in sedimentary rocks. Instead, they are found in igneous rock or rock made from cooled magma, in which fossils cannot form. (Igneous: having solidified from lava or magma; relating to volcanic processes. Lava: molten or semi-fluid rock erupted from a volcano, or solid rock resulting from cooling of this. Magma: hot fluid or semi-fluid material within the Earth's crust from which lava and other igneous rock is formed by cooling.[1])

It is therefore necessary, said science writer Tracy V. Wilson, for researchers 'to find a neighbouring layer of earth that contains igneous rock such as volcanic ash. These layers are like bookends – they give a beginning and end to the period of time when the sedimentary rock formed.'[2]

The dating of a fossilised bone itself, therefore, cannot be performed directly *per se*, and this method of radioactive dating gives only an approximate guide, and is not an exact science.

Chapter 12

How Life Began, How Living Creatures Evolved, and the Dinosaurs in Particular

Naturalist Charles Robert Darwin (1809–82), co-originator (with naturalist, explorer, geographer, anthropologist and biologist Alfred Russel Wallace (1823–1913) of the theory of evolution by natural selection, pondered over the question as to how life originally began. (Natural selection: the evolutionary process whereby organisms that are better adapted to their environment tend to survive and produce more offspring.[1])

In a letter that Darwin wrote to botanist Joseph Hooker on 1 February 1871, he stated:

> It is often said that all the conditions for the first production of a living organism are now present, which could ever have been present. But if (& oh what a big if) we could conceive in some warm little pond with all sorts of ammonia & phosphoric salts, light, heat, electricity, &c., present, that a protein compound was chemically formed, ready to undergo still more complex changes, [then] at the present day such matter would be instantly devoured or absorbed, which would not have been the case before living creatures were formed.[2]

In other words, the chances of such an event happening were extremely remote.

Nevertheless, Darwin had once said to Wallace, in August 1872, that 'I should like to see Archebiosis proved, if it could be shown that life had generated itself spontaneously, then this would be a discovery of transcendent importance.'[3] (Archebiosis is defined as abiogenesis, especially as relating to the origin of life on Earth. Abiogeneis: spontaneous generation.[4])

In 1953, Harold C. Urey of the University of Chicago and his 23-year-old graduate student, Stanley L. Miller, conducted an experiment in which they simulated conditions believed to be present at the time life on

Earth began. In the experiment, electric sparks (i.e. simulating lightning) were continually passed through a flask containing water (heated), methane, ammonia and hydrogen (but not oxygen, as there was little or none of this element/gas in the atmosphere before plant life began). Two weeks later, Miller and Urey observed that 2 per cent of the carbon present in the flask was now in the form of amino acids – organic compounds that occur naturally in plant and animal tissues and which are the basic constituents of proteins.

US Professor Carl Sagan (1934–96), astronomer and astrophysicist, subsequently conducted a similar experiment in which 'a rich collection of complex organic molecules, including the building blocks of the proteins and the nucleic acids' was created. (Nucleic acid: a complex organic substance, especially DNA or RNA. DNA: a substance present in nearly all living organisms as the carrier of genetic information. RNA: a substance in living cells involved in the synthesis of proteins encoded by genes and in some viruses carrying genetic information instead of DNA. Gene: a unit of heredity that is transferred from a parent to offspring and is held to determine some characteristic of the offspring.[5])

Sagan observed that 'Under the right conditions, these building blocks assemble themselves into molecules resembling little proteins and little nucleic acids. These nucleic acids can even make identical copies of themselves.'[6]

This research gives an indication that in all probability, life on Earth began spontaneously, a phenomenon called 'abiogenesis'.

On 30 June 1858, Darwin and Wallace agreed that papers be deposited at the Linnean Society (a biological society for the study and promotion of all aspects of the biological sciences), London, containing 'the results of the investigations of two indefatigable naturalists' – i.e. themselves. The papers were accompanied by a letter, which read as follows:

> These gentlemen having, independently and unknown to one another, conceived the very same very ingenious theory to account for the appearance and perpetuation of varieties and of specific forms on our planet, may both fairly claim the merit of being original thinkers in this important line of inquiry.

Darwin and Wallace, who had each come to his conclusion independently, had observed how living things vary in form and nature, or 'mutate'. If such variation is favourable to survival, those with the most favourable mutations are most likely to survive and reproduce, whereas those with disadvantageous mutations will tend to die out. In other words, only the 'fittest [i.e. those most suitably adapted to their environment] would survive'.[7] Furthermore, it is likely that the offspring of the survivors will carry beneficial features, encoded in their genes, and they will probably transmit these features to the next generation, hence evolution by natural selection.

In August 1858, Wallace's essay, entitled 'On the Tendency of Varieties to Depart Indefinitely from the Original Type', and an abstract of Darwin's manuscript work on 'Species' were duly published in the *Journal of the Proceedings of the Linnean Society*.[8] However, a disappointed Darwin said in his autobiography, 'Our joint productions excited very little attention.'[9]

What Darwin and Wallace had proposed was a theory of evolution whereby every plant and animal on the planet had evolved from more primitive forms as a result of minor changes (mutations) in their structure on the one hand and natural selection on the other. Furthermore, in his *First Notebook*, entitled 'Transmutation of Species' (1837), Darwin sketched out the very first rudimentary diagram of an evolutionary tree of life.

What was the lineage of the dinosaurs? As already mentioned, according to the Cambridge University/Natural History Museum (CU/NHM) study of 2017, the first dinosaurs appeared about 247 mya – i.e. in the Early Triassic Period.

Dinosaurs are members of the class Reptilia, group Archosauria. This group includes two major subgroups: Avemetatarsalia, which includes dinosaurs, pterosaurs (extinct flying reptiles) and birds, and Pseudosuchia, which includes crocodilians and their extinct relatives. In other words, it was in the Early Triassic Period that dinosaurs evolved as a breakaway clade ('Dinosauria') within the group Archosauria, subgroup Avemetatarsalia.

Chapter 13

The KT Boundary and the KT Extinction Event

Between the end of the Cretaceous Period (66 mya) and the beginning of the Palaeogene Period (66–23 mya), the Earth's flora and fauna underwent a remarkable and drastic transformation, in which an estimated 80 per cent of all species became extinct.

Although the KT boundary represents but a small blip on the geological timescale, it is hugely significant biologically: 'Almost all the large vertebrates on Earth, on land, at sea, and in the air, suddenly became extinct.'[1] Why did the KT extinction event impact disproportionately on the Earth's larger creatures. This was because such creatures were at the top of a food chain that was collapsing from the bottom up, with the extinction of flora on both land and in the oceans (i.e. plants and plankton, etc., respectively) and being of great size, they therefore required more food than was available. (Plankton: the small and microscopic organisms drifting or floating in the sea or fresh water, consisting chiefly of diatoms, protozoans, small crustaceans, and the eggs and larval stages of larger animals.[2] Diatom: a single-celled alga that has a cell wall of silica. Protozoan: a single-celled microscopic animal of a group of phyla of the kingdom Protista, which includes amoebas, flagellates, ciliates and sporozoans.[3])

On Land

Said Sankar Chatterjee, Paul Whitfield Horn Professor of Geosciences and Curator of Paleontology at Texas Tech University, Lubbock, Texas, USA, 'all land animals weighing more than 25 kilograms [55 pounds] disappeared from the planet'.

On land, many species of marsupial mammal became extinct, and many species of reptile were depleted or destroyed. Of the creatures that could fly, the reptilian pterosaurs, and several species of birds, became extinct.[4] Also, said Chatterjee, hundreds of plant species were 'suddenly wiped out'.[5]

In the Oceans

'Large marine reptiles, such as plesiosaurs and mosasaurs became extinct.' Their cousins, the ichthyosaurs, are believed to have already become extinct, about 95 mya, in the Cenomanian Age: 100.5–93.9 mya of the Late Cretaceous period – i.e. well before the KT extinction. The same applies to the largest of the marine turtles.

There was severe extinction of major reef-building bivalves called rudists, spiral-shaped ammonites, and their close relatives belemnites; many species of brachiopods, echinoids and fish as well as two-thirds of all marine species were also extinguished.[6] (Bivalve: an aquatic mollusc that has a compressed body enclosed within a hinged shell, such as an oyster, mussel or scallop. Rudist: a cone-shaped fossil bivalve mollusc that formed colonies resembling reefs in the Cretaceous Period. Mollusc: an invertebrate of a large phylum [Mollusca] including snails, slugs and mussels, with a soft unsegmented body and often an external shell. Belemnite: an extinct marine cephalopod mollusc with a bullet-shaped internal shell, typically found as fossils in deposits of the Jurassic and Cretaceous Periods. Cephalopod: an active predatory mollusc of the large class Cephalopoda, which comprises octopuses, squids and cuttlefish. Brachiopod: a marine invertebrate of the phylum Brachiopoda, which comprises the lamp shells. Echinoderm [echinoids]: a marine invertebrate of the phylum Echinodermata, which includes star fishes, sea urchins, brittle-stars, crinoids and sea cucumbers.[7])

There was also 'severe extinction of planktonic foraminifera'.[8] (Foraminifer: a single-celled planktonic animal with a perforated chalky shell through which slender protrusions of protoplasm extend.[9])

The Cretaceous/Palaeogene (KT) boundary is marked by a distinct geological stratum, which occurs globally. This is no coincidence. In fact, the KT boundary, which is never more than a few feet in thickness at the most, holds the key to understanding the great extinction that occurred at this time.

The KT boundary marks the boundary between the Maastrichtian Age (72.1–66 mya: the final age of the Late Cretaceous Period) and the Danian Age (66–61.6 mya: the first age of the succeeding Palaeogene Period). (The 'K' in 'KT' stands for 'Cretaceous' and is derived from the German word 'Kreide', meaning 'chalk', and the 'T' stands for 'Tertiary' – i.e. the Tertiary Period (66–2.6 mya). However, the name Tertiary has fallen into disuse, and therefore the term 'K-Pg boundary' is more accurate, 'Pg' standing for 'Palaeogene'. Nonetheless, KT is preferred in this narrative.)

Geologically speaking, the KT boundary is defined by a layer of dark-coloured rock, up to a few feet in thickness at the most, that is present not only on land but also beneath the sea.

The Nature of the KT Boundary

KT boundary segments throughout the world are characterised by the following constituents:

Shocked Quartz

Shocked quartz is formed when high-pressure shockwaves pass through quartz causing shock lamellae to form. The formation of shocked quartz 'is a distinctive signature of an impact event, as it can only form at a force of more than 10 gigapascals [when it] travels through the quartz-bearing grains of the target rock to produce microscopic shock lamellae'.[10] (Quartz: a hard mineral consisting of silica, found widely in igneous and metamorphic rocks. Igneous: having solidified from lava or magma. Metamorphic: rock that has undergone transformation by heat, pressure, or other natural agencies. Lamella: a thin layer or membrane. Pascal: the SI unit of pressure, equal to one Newton of pressure per square metre.[11] SI [Système Internationale]: the international system of measurements based on the metric system. Target rock: i.e. the rock to which the force is applied.)

An asteroid is defined as 'a small, rocky body orbiting the sun', whereas a meteorite is defined as 'a piece of rock or metal that has fallen to the earth as a meteor'.[12,13] Therefore, to avoid confusion, the word 'asteroid' will be used in this narrative.

Spherules

Glassy droplets a few tenths of a millimetre in diameter of felsic (relating to or denoting a group of light-coloured minerals including feldspar, quartz and muscovite[14]). In this context, droplets of molten rock.

Stishovite

A dense form of silica produced by very high pressures.[15]

Soot

A black powdery or flaky substance consisting largely of amorphous carbon, produced by the incomplete burning of organic matter. (Amorphous: without clearly defined shape or form.[16])

Is it certain that no Dinosaurs Whatsoever Survived the KT Extinction?

If it is the case that dinosaurs became extinct during the period that the KT boundary rock formation was laid down, then one would expect fossilised dinosaur bones to be found both within and below the boundary itself, but not above it. Until recently, the nearest dinosaur fossil to be discovered to the KT boundary was the brow horn from a dinosaur of the family Ceratopsidae, probably an adult of the genus *Triceratops*. Such creatures grew to a length of 30 feet and weighed up to 5 tons. This fossil was discovered 5 inches below the KT boundary – i.e. it was from an earlier time.[17]

Surprisingly, dinosaur fossils have been found above the KT boundary, corresponding, said Dan O'Dea, Science Educator at the Children's Museum of Connecticut, to 'as much as a million years after the date of the boundary'. But these fossils were all 'small, individual parts: a broken horn or two, scattered teeth, and occasional scraps of bone'. How may this be explained?

O'Dea continued: 'Current theory ... indicates that these are reworked deposits; that is, each fossil found [had been previously] eroded from some Cretaceous layer and was deposited, along with other bits of the earlier rocks, in younger sedimentary deposits.'[18]

Meanwhile, in New Jersey, at a quarry the site of which once lay beneath the sea, Dr Kenneth Lacovara, director, Edelman Fossil Park of Rowan University, Glassboro, said of the KT boundary: 'The animals found here are typical of the Late Cretaceous ... but no one in the world has found an in-place dinosaur [as much as] 1 centimeter above this line' – i.e. above the top of the KT boundary.[19]

Is the KT boundary a truly global phenomenon? Evidently so, for the sites where it has so far been discovered – Italy, Denmark, Spain, Austria, Tunisia, Turkmenistan, New Mexico USA, and New Zealand – are widely distributed.

Clearly, extremely high temperature and pressure were required to create the KT boundary. So how may this be explained?

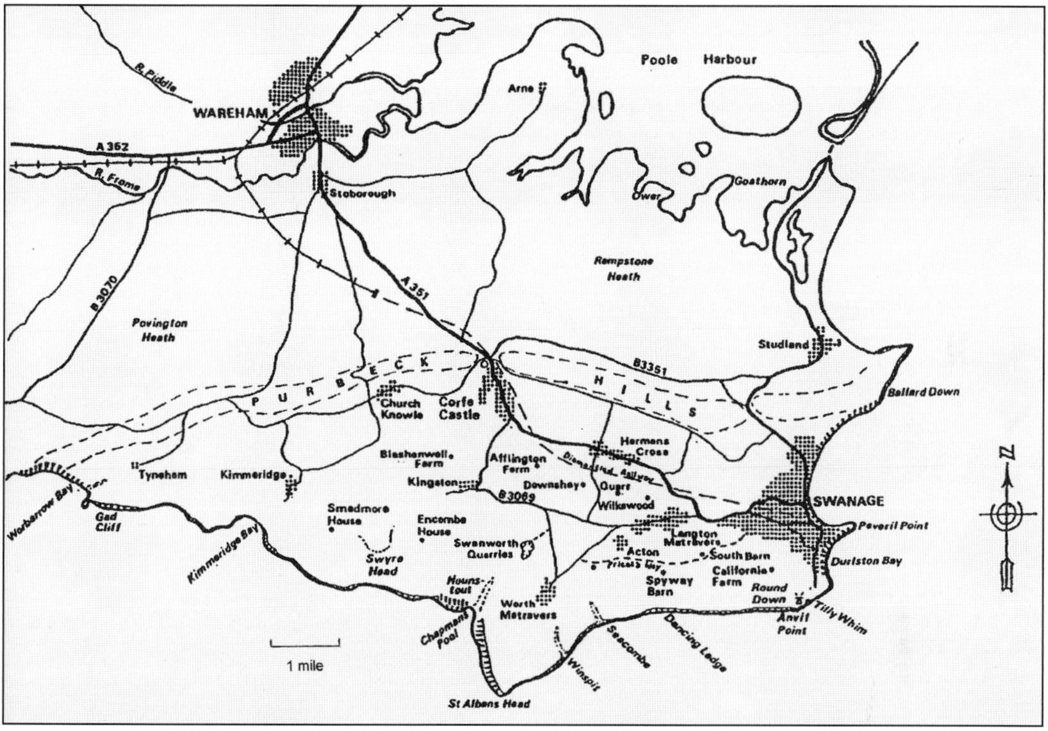

Isle of Purbeck. (*Geoffrey Norris*)

Swanage's guardian dinosaur! (*Santa-Fe Fun Park, Swanage*)

Ammonite (32in × 28in), built into the wall of the Tithe Barn, Swanage.

William Buckland. Mezzotint by S. Cousins, 1833, after T. Phillips. (*Wellcome Collection*)

'The Extinct Animals': Benjamin Waterhouse Hawkins at work on the iguanodon mould in his model room at Crystal Palace Park, Sydenham Hill. (Illustrated London News)

'He was at my very heels: I was lost', illustration by Patrick L. Forbes for Sir Arthur Conan Doyle's novel *The Lost World*. (The Strand *magazine, August 1912*)

Mary Anning by Benjamin Donne (after unknown artist). (*Natural History Museum*)

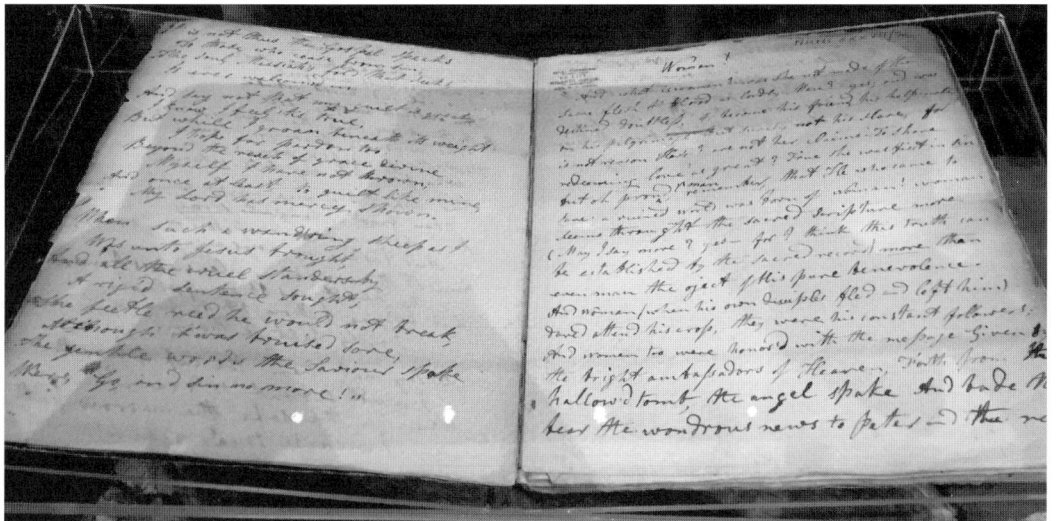

Mary Anning's Commonplace Book. (*Lyme Regis Museum*)

Henry de la Beche. (*Wellcome Collection*)

Glen J. Kuban on the Ozark Site, Dinosaur Valley State Park, Paluxy Riverbed, Glen Rose, Texas: showing striding trails of sauropod tracks (left), made by a large, four-legged plant-eating dinosaur, probably *Sauroposeidon proteles*. The hind prints measure about 1.1 metres (or more than 3½ feet in length). Moving in a similar direction (right), the sharp-clawed, three-toed tracks of the predatory theropod dinosaur *Acrocanthosaurus tokensis*. (*Glen J. Kuban*)

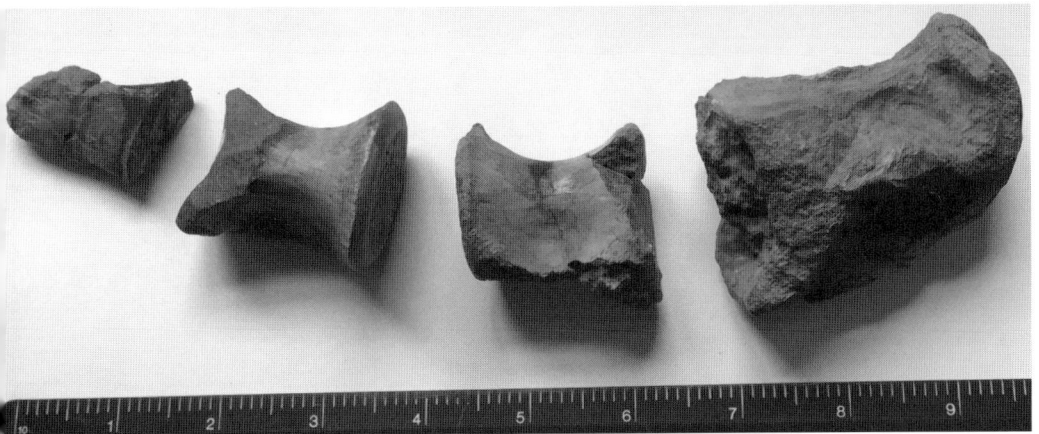

Dinosaur tail vertebrae discovered at Peveril Point, Swanage. (*Swanage Heritage Centre*)

(Left) *Iguanodon* tail vertebra discovered in Wealden beds. (*Swanage Heritage Centre*)
(Right) Cast of dinosaur hind-foot print (17in × 16in), probably that of an *Iguanodon*, found near Swanage. (*Dorset County Museum*)

Jawbone (1½in long) of dinosaur *Nuthetes*. (*Dorset County Museum*)

The Common Poorwill (*Phalaenoptilus nuttalii*). (Connor Long)

The four spinal fossilised vertebrae of *Vectaerovenator inopinatus*. (University of Southampton)

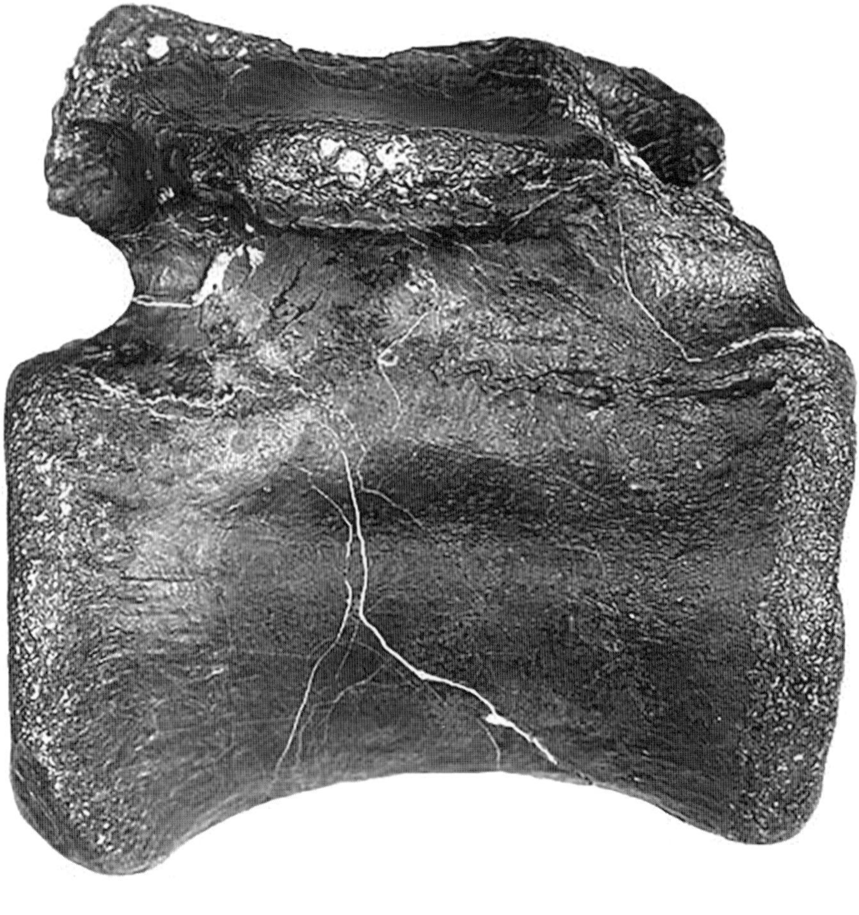

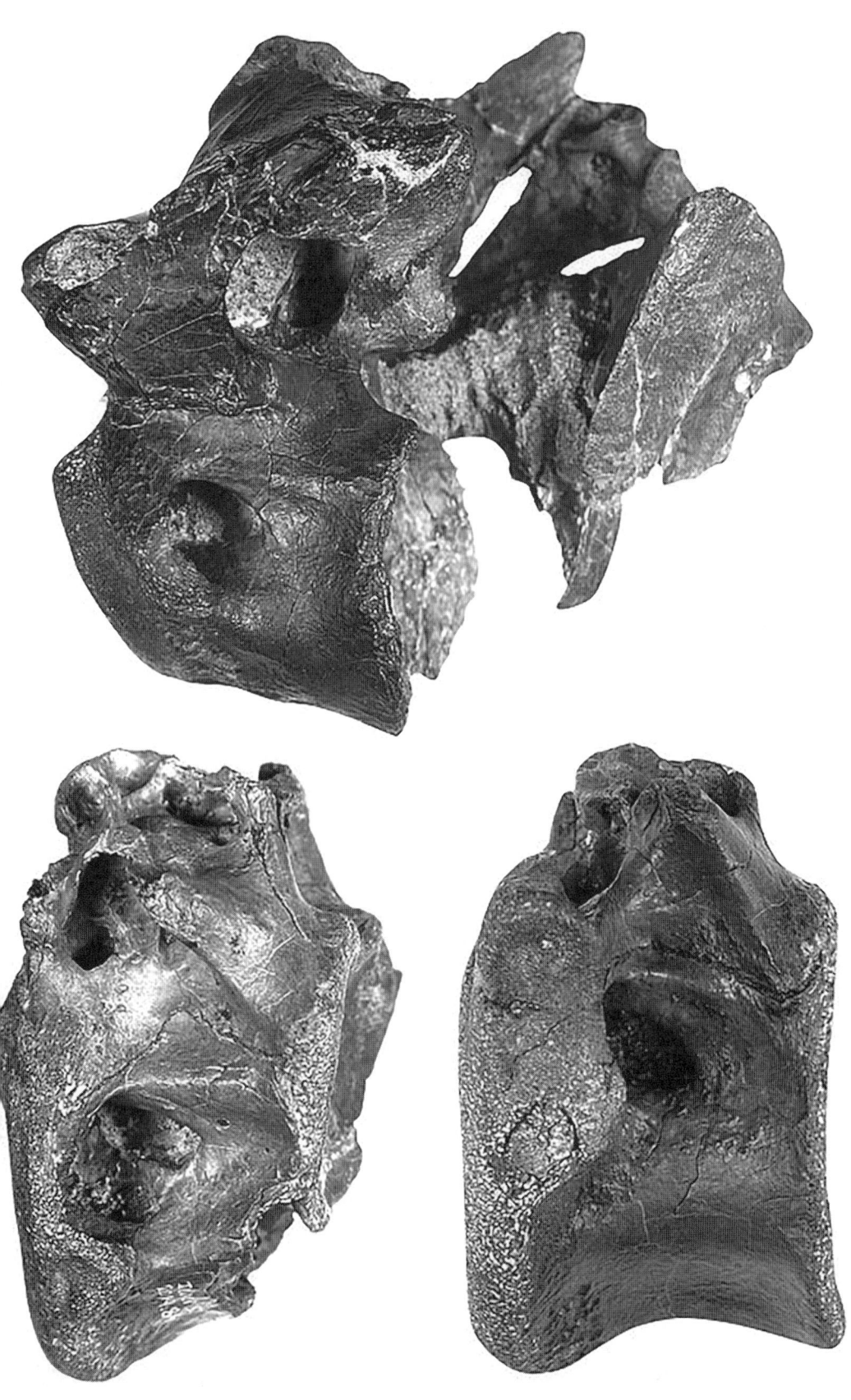

Dinosaur vertebra discovered by Paul I. Farrell on Shanklin beach. (*Paul Farrell*)

Dinosaur Isle Museum, Sandown, Isle of Wight. (*Martin Munt*)

Life-sized cast of the sauropod dinosaur *Patagotitan* from Patagonia, Argentina. (*Natural History Museum*)

Patagotitan, 55–63 tons in weight and 102 feet in length, tail extending into an adjacent room! (*Natural History Museum*)

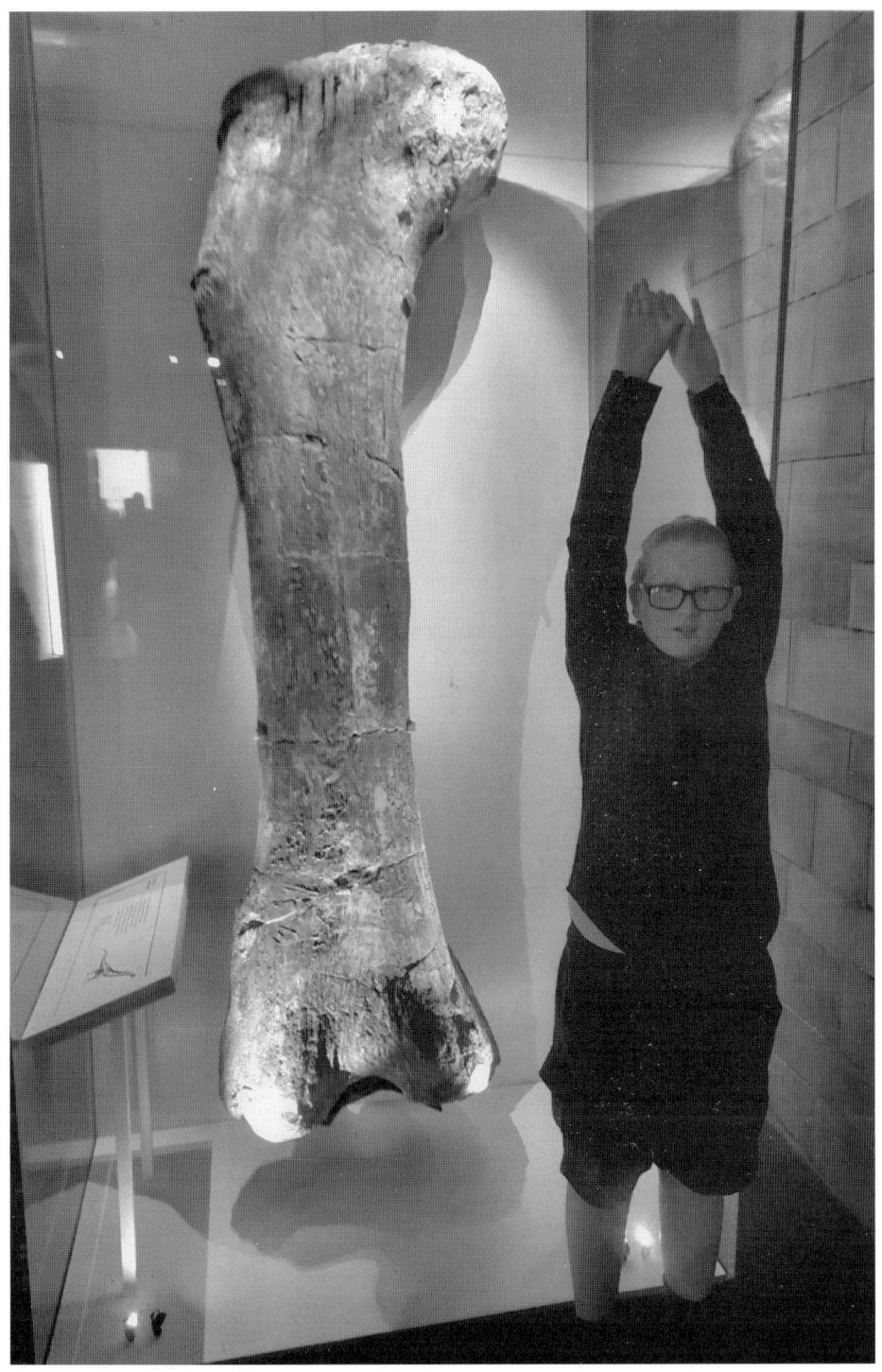

Patagotitan: fossilised 7 feet 10 inch long femur (thigh bone), with Austin Dodd, aged 11. (*Natural History Museum*)

Patagotitan, with three dinosaur enthusiasts: Gabriella, Gabriel and Olivia Dragffy!
(*Natural History Museum*)

Chapter 14

The Presence of High Concentrations of Iridium at the KT Boundary, and its Significance

In June 1980, Luis W. Alvarez, Professor Emeritus of Physics at Lawrence Berkeley Laboratory, University of California, Berkeley, USA, and his geologist son, Walter, together with nuclear chemists Frank Asaro and Helen V. Michel, published the results of a study, entitled 'Extraterrestrial Cause for the Cretaceous-Tertiary Extinction'.[1] He stated: 'In this article, we present direct physical evidence for an unusual event at exactly the time of the extinctions in the planktonic realm.'[2]

Alvarez continued: 'This study began with the realization that the platinum group elements [which include iridium] are much less abundant in the Earth's crust and upper mantle than they are in chondritic meteorites and average solar system material.'[3] (Chondrite: a stony meteorite containing small mineral granules.[4])

Why is a greater concentration of iridium to be found in an asteroid than in the upper strata of the earth? After all, both are derived from the same primordial proto-planetary soup (i.e. dense gas and dust that existed before the planets were formed and out of which they were made). However, the Earth's iridium is believed to have descended towards its core when the earth was still molten.

Alvarez *et al.* now decided 'to measure the iridium concentration in the 1-centimeter-thick clay layer that marks the C-T [KT] boundary in some sections in the Umbrian Apennines [mountains of Italy]', the 'best known' of these sections being 'in the Bottaccione Gorge near [the town of] Gubbio'.[5]

Alvarez discovered that the level of iridium 'increases by a factor of about 30 in coincidence with [i.e. at] the C-T [KT] boundary', in comparison with twenty-seven other elements tested.[6]

Further experiments were performed at KT boundary (KTB) sites in Denmark, where 'the Ir [iridium] in the boundary layer residue rises by about a factor of 160 over the background level'.[7] The reasons for this factor-of-ten difference in the iridium content of the boundary clay between Denmark and Italy are, as yet, not fully understood.[8]

Whereupon, the team concluded, in the case of both the Italian and the Danish levels of iridium at the KTB, that they were most unlikely to 'have had a crustal origin' – i.e. did not originate from the Earth's crust.[9] Furthermore, at the time of writing (June 1980), the 'iridium anomaly' had also been detected at 'two different areas in Western Europe and in New Zealand' – i.e. the iridium anomaly was evidently a global phenomenon.[10]

The team concluded, therefore, that 'the anomalous iridium concentration at the C-T boundary is best interpreted as indicating an abnormal influx of extraterrestrial material' – i.e. material originating from outside the Earth, or its atmosphere.[11]

Alvarez *et al.* now proposed the following hypothesis: '... that an asteroid struck the Earth, formed an impact crater, and some of the dust-sized material injected from the crater reached the stratosphere and was spread around the globe'. (Stratosphere: the layer of the earth's atmosphere above the troposphere extending to about 50 kilometres [about 31 miles] above the earth surface. Troposphere: the lowest region of the atmosphere, extending from the earth surface to a height of about 6–10 kilometres [3.7–6 miles].[12]) This dust effectively prevented sunlight from reaching the surface [of the Earth] for several years, until the dust settled to Earth. Loss of sunlight suppressed photosynthesis, and as a result most food chains collapsed, and the extinctions resulted.[13]

Walter Alvarez stated:

> We have very strong physical and chemical evidence for a large impact and this is the most firmly established part of the whole story. There is unquestionable mass extinction at this time and in the fossil groups for which we have the best record, the extinction coincides with the impact to a precision of a centimetre or better in the stratigraphic record.

(Stratigraphy: the branch of geology concerned with the order and relative dating of strata.[14])

'This exact coincidence in timing argues for a causal relationship.'[15] In other words, it was the asteroid impact that caused the mass extinction. Such an asteroid 'would enter the atmosphere roughly 25 km (15 miles) per second [i.e. 54,000 mph] and would punch a hole in the atmosphere about 10 km (6 miles) across. The kinetic energy [energy that a body possesses by virtue of being in motion[16]] of the asteroid is approximately equivalent to that of 10^8 megatons of TNT.'[17]

In support of the asteroid impact theory is that 'meteorite fragments have been described from the KT boundary layer clays in the North Pacific Ocean that may represent samples of a KT asteroid impactor'. Said geochemist Frank T. Kyte, of the University of California, Los Angeles: 'Here we describe a 2.5 mm fossil meteorite found in sediments retrieved from the Cretaceous/Tertiary boundary in the North Pacific Ocean that we infer may be a piece of the projectile responsible for the Chicxulub crater.' (More will be said about this crater shortly.) Furthermore, 'geochemical and petrographic analysis' confirmed that the crater had been created by 'a typical metal-and-sulphide-rich carbonaceous chondrite, rather than from cometary materials'.[18] (Petrography: the study of the composition and properties of rocks. Chondrite: a stony meteorite containing small mineral granules. Comet: a celestial object that consists of a nucleus of ice and dust and, when near the sun, a diffuse tail, and typically follows a highly eccentric orbit around the sun. Carbonaceous: consisting of or containing carbon or its compounds.[19])

To account for the quantity of iridium deposited on the surface of the Earth (including the oceans), the team calculated that the diameter of the impacting asteroid would be 'about $10\pm 4\,km$' – i.e. between 3.7 and 8.7 miles, or just over 6 miles in diameter on average. Furthermore, the team estimated that the diameter of the impact crater would be in excess of 100km (60 miles).[20]

Finally, said Alvarez *et al.*: '... we would like to find the crater produced by the impacting object'.

Chapter 15

Chicxulub: A Possible Location for the Alleged Asteroid Impact

The surface of the moon gives an indication of the devastation that asteroids can cause, for its surface is pitted with craters: the largest being the Aitken lunar impact crater (or Aitken basin), which measures a colossal 1,600 miles or so in diameter. Following Luis Alvarez *et al.* and his suggestion of a meteorite impact to account for the iridium found in the KT boundary rock layer, the question is, can an earthly asteroid-induced crater be identified that fits the bill?

In 1978, Glen Penfield and Antonio Camargo of Pemex, the Mexican National Petroleum Company, discovered a possible candidate while prospecting for oil in Mexico. They were conducting a survey to the north of the Yucatán Peninsula in the Gulf of Mexico when they discovered a huge crater, half on land and half beneath the sea. The crater is buried beneath 2,000 feet or so of limestone and other sediment that has accumulated in the 66 million years since the impact. It was subsequently named after the nearby Mayan village of Chicxulub.

In 1991, Alan Hildebrand and William Boynton of the Canadian Geological Survey showed that the Chicxulub Crater had the 'typical morphology of a complex crater with a central peak, surrounded by an annular basin and a faulted outer rim'.[1]

Is it possible to link the creation of the Chicxulub Crater with the alleged asteroid impact timewise – i.e. to prove that the two events occurred simultaneously?

In 2013, Paul R. Renne, Professor in Residence of Geology, Department of Earth & Planetary Science, University of California, Berkeley *et al.*, announced the results of a study in which they attempted to answer this question. Their method was to compare isotopes in tektites from the Chicxulub event with the same isotopes in bentonites 'only a few centimetres above the largest iridium anomaly' – i.e. immediately above the KT boundary (KTB) rock layer. (Tektites: small, black, glassy objects, or 'spherules', found in numbers over certain parts of the Earth's surface, believed to have been formed as molten debris in meteorite impacts and scattered widely through the air. Bentonite: a kind of absorbent clay formed by breakdown of volcanic ash.[2])

The age of the bentonites, produced by volcanic activity in the vicinity of the KTB, would give an approximation as to the age of the adjacent rock stratum.

> We analysed multiple samples of the tektites to refine [i.e. define more precisely] the age of the Chicxulub impact, and of bentonites (altered volcanic ashes) clearly associated with the KTB. We analysed samples from two localities about 200 miles apart [in north-eastern Montana] of a bentonite ... located stratigraphically only a few centimetres above the horizon [level] yielding the largest iridium anomaly [i.e. immediately adjacent to the KTB rock layer].

The outcome was that the respective dates, 66.038 ± 0.049 mya (for the tektites) and 66.019 ± 0.021 mya (for the bentonites) were within 19,000 years of each other – i.e. identical within the limits of experimental error.[3]

In March 2010, in the *National Geographic* magazine, science writer David Braun, in an article entitled 'Asteroid Terminated Dinosaur Era in a Matter of Days', described how 'a panel of 41 experts from Europe, the US, Mexico, Canada and Japan' (including from Imperial College, London, and from the University of Texas at Austin) had 'analyzed new data from ocean drilling and continental sites and reviewed the research of palaeontologists, geochemists, climate modelers, geophysicists and sedimentologists who have been collecting evidence over the last 20 years to determine the cause of the Cretaceous-Tertiary (KT) Extinction, which happened around 65 [actually 66] million years ago'.

The team was led by Peter Schulte, geophysicist at the University of Erlangen-Nurnberg, Nuremberg, Germany, and their findings were published in the journal *Science* on 5 March 2010, under the heading 'The Chicxulub Asteroid Impact and Mass Extinction at the Cretaceous-Paleogene boundary'.

The conclusion of Schulte and his team, in respect of the asteroid impact that Alvarez *et al.* had originally postulated, was as follows: 'That the extinction was caused by a massive asteroid slamming into the Earth at Chicxulub in Mexico'; that the asteroid was 'around 15 kilometers [9 miles] wide' (which approximately accords with Luis Alvarez *et al.*'s upper estimate for the asteroid's diameter of 8.7 miles); and that, 'At impact, the asteroid is estimated to have been traveling at 20 kilometers per second (44,640 mph, roughly 20 times the speed of a rifle bullet).' It is important to note that the asteroid possessed potential energy, on account of its mass and velocity, but it also possessed kinetic energy, on account of the heat generated in it by its entry into the gas-containing Earth's atmosphere. Images of modern-day spacecraft show their heat shields glowing red with the enormous friction generated.

The US spacecraft Orion, for example, re-enters the atmosphere at a speed of 24,480 mph, whereupon its surface reaches a temperature of 2,649 degrees Celsius.[4] The asteroid, however, is estimated to have weighed between 1 and 2 trillion tons and was, therefore, at least 100 billion times more massive than Orion (9.8 tons), and it entered the atmosphere at almost twice the speed (i.e. at about 45,000 mph). (Luis Alvarez *et al.* had proposed 54,000 mph.)

Continued the experts: 'The initial impact crater was about 100 kilometers (60 miles) wide [Alvarez *et al.* had proposed in excess of 100 km] and 30 kilometers (18 miles) deep.' 'Impacts of this size on Earth are thought to happen on average about once every 100 million years.'

> Some scientists have suggested that the Chicxulub impact happened 300,000 years prior to the KT boundary came into being, and therefore came too early to have been the major cause of extinctions. [They] point to deposits at sites around the Gulf of Mexico with a layer of tiny glass-like blobs of melted impact material that, according to their interpretation, was deposited at about 300,000 years before the mass extinction.
>
> However, 'the reviewers of the study published today find that what appears to be a series of layers neatly laid down over 300,000 years near the impact site, were actually violently churned and then dumped in a thick pile in a very short time'. In other words, the impact material had been laid down suddenly, following a violent upheaval (which is consistent with an asteroid impact).

Models suggest the impact at Chicxulub was a million times more energetic than the largest nuclear bomb ever tested. An impact of this size would eject material at high velocity around the world, causing earthquakes of a magnitude >10 [on the Richter Scale]; Continental Shelf collapse; landslides; gravity flows [flows of water or liquid drawn under the influence of gravity]; mass wasting [large movements of rock, soil and debris downwards due to the force of gravity]; tsunamis; and produce a relatively thick and complex sequence of deposits close to Chicxulub.

Chicxulub: A Possible Location for the Alleged Asteroid Impact

In addition, the reviewers note, as you go farther from the impact site, these layers become thinner and the amount of ejected material decreases, until it becomes one layer that can be found globally exactly at the KT boundary coincident with the mass extinction. Moreover, the ejecta (blasted material) within the global KT layer is compositionally linked to the specific sediments and crystalline rocks at Chicxulub.[5]

In other words, the ejecta (material that is forced or thrown out as a result of meteoritic impact[6]) bears the geological signature of Chicxulub.

Finally, as already pointed out, there was an uncanny similarity between the predictions of Luis Alvarez *et al.* (based on the finding of iridium in the KT boundary). On the one hand, in respect of the size of the asteroid, its speed of entry into the Earth's atmosphere, and the size of the crater produced (at Chicxulub); and on the other hand, the conclusions of the panel of forty-one international experts.

Chapter 16

Was there more than one Asteroid Impact?

Could the impact of a single asteroid have caused the observed catastrophic changes, both locally and globally, to the Earth's ecology, or was there more than one?

Frederick L. Sutherland, Senior Fellow Geologist at the Australian Museum, stated as follows: 'Upper estimates of the effects of acid rain and temperature changes are excessive when considered in the light of a single impact event, such as Chicxulub, but are consistent with indications for [i.e. correlate with] multiple KTB [KT boundary] impacts.'[1]

Could it therefore be that the Chicxulub asteroid impact was only one of two or more that occurred at about the same time? The answer is a possible yes.

The Shiva Crater

Sankar Chatterjee identified another possible candidate: 'I discovered and named the Shiva crater around Bombay High [an offshore oilfield 30 miles off the coast of Mumbai, India] in 1992.' The crater is in the Arabian Sea, 'in the Mumbai Offshore Basin on the western continental shelf of India'. It is 'largely submerged' beneath the waters of the Indian Ocean, and buried by a 1.2–4.3-mile-thick strata of 'Tertiary sediments'.[2]

Chatterjee continued: 'The 500 kilometre [310 mile] diameter [of the] Shiva crater suggests an impact projectile of 40 kilometres [25 miles] in diameter.'[3] The Shiva impact, therefore, clearly dwarfed the Chicxulub impact in magnitude: the Chicxulub crater being an estimated 60 miles in diameter, having been caused by an asteroid an estimated 9 miles in diameter.

Furthermore, whereas 'the pressure exerted by the Chicxulub impact would have generated more than a 100 million-megaton blast, the Shiva impact would have been at least 10 times more catastrophic'.[4]

In respect of the Shiva impact, Chatterjee pointed out that the 'target rocks' (i.e. those with which the Shiva asteroid impacted) are composed of

Proterozoic granite (Proterozoic Eon: 2,500–542 mya) and Deccan basalt, both of which are 'poor in iridium content'.[5] But referring to 'impact melt rocks' discovered both inside and outside the crater, Chatterjee quoted Anil D. Shukla of the Physical Research Laboratory, Ahmedabat, India, when he stated that these rocks are 'rich in iridium because of contamination from impacting meteorites [i.e. more than one meteorite]'.[6]

In addition, said Chatterjee, at various KTB sections around the Shiva crater, 'the iridium anomaly, shocked quartz, and spherules have been reported'.[7] This is the typical outcome of an asteroid impact, as already mentioned.

With the impact of the asteroid, the rock formations at Shiva (as at Chicxulub) would have instantaneously vaporised; the ejecta reaching high altitude where it cooled and condensed into droplets (spherules), which then fell to Earth.

Similarly, 'an anomalously high concentration of iridium has been found in two KT boundary sections of India that lie beyond the Deccan Province': one at Meghalaya and the other at Anjar in Gujarat.[8]

Apart from the local effects produced by the impact, which were considerable, Chatterjee describes the 'distal ejecta' – i.e. that which was hurled into the atmosphere and thence dispersed over a wide area. This included 'shocked quartz, iridium anomalies, highly magnetic nano-particles, fullerenes, magnetic spherules, nickel-rich spinels, and Deccan basalt spherules'.[9] (Nano: submicroscopic. Fullerene: a form of carbon having a molecule consisting of a large spheroidal cage of atoms, produced chiefly by the action of arc [electrical] discharges between carbon electrodes. [How these fullerenes were formed is not fully understood.] Spinel: a hard, glassy mineral consisting chiefly of magnesium and aluminium oxides.[10])

Chatterjee continued: 'Since the target rock of the Shiva impact includes the Deccan basalts, the basaltic spherules recovered from different KT boundary sections at several locations around half the globe must have come from the Shiva impact.' Chatterjee also observed that 'the Chicxulub target rock, in contrast to [that of] Shiva, does not contain any basaltic layer and could not be the provenance for the global distribution of basaltic spherules'.

Conclusion

It was Chatterjee's opinion that the Chicxulub and Shiva craters were caused by 'twin asteroid impacts'. This, in turn, implies that the asteroids involved were derived from fragments of a larger asteroid. Is this a possibility? Yes. For example, in September 2015, twin craters were discovered in Sweden, each in exactly the same layer of sediment, indicating that the two impacts had occurred simultaneously. In July 2018, NASA discovered a twin asteroid in space, each of its components being about half a mile in diameter.[11] It was given the name '2017YE5'.

Before the discovery of the Shiva impact site, some have questioned whether the Chicxulub asteroid could have contained sufficient iridium to cover the entire surface of the globe. But add the quantity of iridium in the Chicxulub asteroid to the quantity of iridium in the Shiva asteroid (which was almost thirty times larger), and the notion of twin-asteroid impacts becomes more plausible (assuming that the concentrations of iridium were similar in the two asteroids). (Estimated volume of the Chicxulub asteroid is 3,600 cubic miles. Estimated volume of the Shiva asteroid is 96,000 cubic miles, based on diameters of 60 and 310 miles, respectively.)

Chapter 17

Oil: Another Potentially Lethal Ingredient

Chicxulub

Kunio Kaiho, geochemist, Department of Earth Science, Tohoku University, Sendai, Japan, *et al.* said that the impact of the asteroid with the geological strata of Chicxulub would have sent a huge cloud of vapourised rock ejecta into the stratosphere.[1]

However, it should be noted that the Gulf of Mexico is also a rich source of oil. Said William E. Galloway, Professor Emeritus, Institute of Geophysics, University of Texas at Austin, by the early 1990s, 'more than 230 billion barrels of oil equivalent' (abbreviated to 'Bboe') had been discovered in the Gulf of Mexico. Of this, 130 Bboe units (i.e. 56 per cent) were formed in the 'young Cenozoic Era (i.e. after the KT extinction event); 85 Bboe units (37 per cent) in the Cretaceous Period (pre-KT); and 15 Bboe units (7 per cent) in the Jurassic Period (also pre-KT). Therefore, when the asteroid hit, a massive oilfield (created during the Cretaceous and Jurassic Periods) was already in existence (equivalent to 44 per cent of known Gulf oil reserves in 1990).[2]

The Gulf of Mexico oil reserves lie at a depth of between 1,000 and 10,000 feet.[3] Furthermore, at the time of the KTB, these deposits would have been covered with less sediment, and therefore even closer to the surface of the ocean than they are today. The depth of the Chicxulub crater, however, was approximately 18 miles. Clearly, therefore, the asteroid could not have failed to penetrate and ignite the adjacent submerged oilfields, creating an explosive mixture of superheated oil and steam: one of the products of oil combustion being soot.

Crude oil has a flashpoint of 60 degrees Celsius – the flashpoint being the temperature at which a particular organic compound gives off sufficient vapour to ignite in air – whereas the temperature at the site of impact was an estimated 10,000 degrees Celsius.[4,5]

Shiva

Mumbai today is also the centre of a huge oil-producing industry. However, the rocks in which the oil is found date from the Early Eocene Epoch (56 mya) to the Mid-Miocene Epoch (14 mya). The creation of the Mumbai oilfield, therefore, postdates the KT extinction event, so oil could not have been a factor in the Shiva impact event.

Burning Oil and Acid Rain

In 1991, Iraqi military forces deliberately set fire to between 605 and 732 oil wells in neighbouring Kuwait, having invaded that country the previous year. These fires burned between January and April, when some 4–6 million barrels of crude oil and 70–100 million cubic metres of natural gas were ignited.[6]

British scientists sent a research aircraft and helicopters to monitor the smoke rising from the Kuwait oil fires. The findings were as follows:

> The high levels of sulfur dioxide, a main cause of acid rain, that were registered on the British flights appeared to confirm earlier reports from weather specialists who followed the smoke's trajectory by satellite. These specialists said oil fires were causing acidic rain at distances of 1,200 miles from Kuwait, and had reached the Black Sea to the north and Pakistan to the east. Last week, snow stained with oily black soot was found on the Himalayan slopes of Kashmir in Northern India.[7]

Therefore, it seems probable that the soot produced by the burning oil from Chicxulub was likewise dispersed over thousands of square miles of land and sea.

In 1996, John L. Ross of the University of Washington, Seattle, published a paper entitled 'Particle and Gas Emissions from an In Situ Burn of Crude Oil on the Ocean'. The report stated as follows: 'The primary combustion products' of the burning of crude oil are 'CO_2 and water'. In addition, dense black smoke is produced, primarily composed of elemental carbon (soot). For each kilogram of fuel consumed, about 87 grams was released as particles with a diameter less than 3.5 millionth of a metre, 'of which about 66 grams was elemental carbon'. In other words, about 9 per cent of the products of oil combustion was soot or particulate matter.[8]

Release of Toxic Gases

Referring to the *in-situ* combustion of the products of an oil spill, physicist Nir Barnea of the National Oceanic and Atmospheric Administration, Seattle, Washington, USA, stated as follows:

> Most of the oil in in-situ burning will be converted to carbon dioxide and water. Particulates (mostly soot) comprise 10% to 15% of the smoke plume. Small amounts of toxic gases are emitted as well. These include sulfur dioxide, nitrogen dioxide, and carbon monoxide. In addition, small amounts of polynuclear aromatic hydrocarbons are emitted from the fire, mostly as residues attached to the particulates.[9]

A study led by Charles Bardeen and published in August 2017 reveals further details about the effect of soot ejecta (resulting from the Chicxulub asteroid impact) on the Earth's ecosystems, as will shortly be seen.

Chapter 18

The Volcanoes of India's Deccan Region

To what extent, if any, did Deccan volcanism (volcanic activity) contribute to the KTB?

During a period of about 3 million years, there was volcanic activity in the Deccan region of India. In the process, huge quantities of lava, mainly tholeiitic basalt, were spewed out across the plain that now lies beneath the Indian Ocean. (Tholeiit: a silica-rich basaltic rock. Basalt: a dark, fine-grained volcanic rock.[1])

These lava beds are known as the 'Deccan Traps'. This, said Sankar Chatterjee, was 'one of the largest volcanic eruptions in Earth's history and today covers an area of 800,000 square kilometres (approximately 300,000 square miles: about the size of Texas) off west-central India and extends seaward more than 500 kilometres [about 300 miles] beyond the modern coastline'.[2]

The Deccan volcanic lava was not extruded all at once, but intermittently over 3 million years. During periods between volcanic eruptions, sediments accumulated in deposits called 'Intertrappean Beds'.

Chatterjee continued that about 1.2 million cubic kilometres (about 300,000 cubic miles) of lava was extruded (forced out) from these volcanoes, the lava flow reaching a maximum thickness 'of about 3.5 kilometres [2 miles]. The main pulse of volcanism (Phase 2) probably erupted very rapidly during a span of 1 million years around 65 million years ago in fact] at the KT boundary and accounts for 80% of all [Deccan] traps.'[3] In fact, the date of the KTB is now estimated at 66 mya.

The KTB occurs in Intertrappean sediments sandwiched between the Deccan lava flows. This indicates that the boundary was formed during the period of Deccan volcanism, and not before or after it. Furthermore, elevated iridium levels are found only at the Deccan's KTB, and not within the rocks of the Deccan traps themselves, indicating that Deccan volcanic activity was not the source of the iridium in the KTB worldwide.

The Evidence

Did Deccan volcanism adversely affect the dinosaurs and other life?

In the Intertrappean Beds between the layers of lava 'are fluvial [relating to or found in a river[4]] or lacustrine [relating to lakes[5]] deposits of lameta sediments [Lameta: a sedimentary rock formation of the Late Cretaceous] that contain abundant remains of plants, invertebrates, dinosaurs, and dinosaur eggs'.[6] Chatterjee stated:

> Bones and eggshells of dinosaurs such as carnivore abelisaurs [saurischian theropods] and herbivore titanosaurs [saurischian sauropods] were found in intertrappean beds of the Deccan Traps, very close to the KT boundary iridium layer, which suggests that the dinosaur extinction was sudden and right at the boundary.

Furthermore, examination of the 'KT boundary iridium anomaly' at Anjar in northwest India in the State of Gujarat (where the north-western section of the Deccan Traps is located) revealed that 'dinosaurs survived the first few phases of Deccan volcanism, but disappeared precisely at the KT boundary'.[7]

Referring to the lakes that had been formed when the lava flows dammed the local rivers and streams, said Chatterjee, they became 'centers for dinosaur communities and became their favourite nesting sites':

> [However,] we could not detect any evidence of biotic crises in these fossil assemblages during episodic volcanic activity. On the contrary, the fossil evidence indicates that dinosaurs were thriving and reproducing during the recurrent Deccan eruptions, quite unaffected by millions of years of volcanic activity.[8]

The Indian stratigraphic evidence clearly indicates, therefore, that the dinosaurs did not become extinct before the formation of the KTB (which was during the early phase of Deccan eruption), but that they lived right up to it. So even though Deccan volcanism 'certainly affected the local flora and fauna by habitat destruction and pollution', said Chatterjee, 'it had little direct effect when it came to 'the major decimation of terrestrial organisms'.[9]

Conclusion

Deccan volcanism was not the cause of dinosaur extinction, the Deccan volcanoes having been previously active for hundreds of thousands of years without any noticeable effect on the dinosaur population.

Chapter 19

The Chicxulub Impact: Both a Local and a Global Catastrophe

The Chicxulub impact would have created a colossal shockwave, hurricane-force winds, a tsunami and possibly earthquakes. The fireball temperature of the order of 10,000 degrees Celsius would have incinerated all terrestrial flora and fauna within a wide area. But this event also had global implications

In 1988, in a paper entitled 'Global Fire at the Cretaceous-Tertiary boundary', Wendy S. Wolbach, Department of Chemistry, DePaul University, Illinois, USA, *et al.* described an investigation of five KTB sites in Europe and New Zealand. They discovered that KTB clays at these sites were '100–10,000 fold enriched in elemental carbon (mainly soot), which is isotopically uniform and apparently comes from a single global fire. The soot layer coincides with the iridium layer, suggesting that the fire was triggered by meteorite impact and began before the ejecta had settled.'

What had fuelled this fire? The team concluded that 'Much or all of the fuel was biomass (plant and animal material) as indicated by the presence of retene (a hydrocarbon diagnostic of resinous wood fires) and by the carbon isotopic composition, which resembles that of natural charcoal and atmospheric carbon particles from biomass fires.' Furthermore, 'the mean amount of elemental carbon at eleven K/T boundary sites' was found to be 'very large and requires that much of the Cretaceous biomass [i.e. the Cretaceous forests and other plant life] burned down and yielded a larger mass fraction of soot and charcoal than small fires'.

The team stated that, following the asteroid impact, 'Initial disruption came from energy dissipated by the impact blast, levelling trees within a radius of 1,500 km, and as intense radiated heat, which may have ignited wildfires on a global scale.'

There have been suggestions that the level of oxygen in the atmosphere was higher at the end of the Cretaceous Period than it is today, in which case the global biomass would have burned more readily. However, according to a 2007 study by Robert A. Berner *et al.*, the oxygen level was in fact lower at that time.[1]

Subsequent Consequences of the Soot Injection

The asteroid impact 'was most likely followed by acid rain resulting from the emission of sulfate-rich vapor and ejection of a large quantity of soot into the atmosphere'.

Following injection into the atmosphere, the soot is heated by sunlight and lofted to great heights, resulting in a worldwide, soot aerosol layer that lasts several years. As a result, little or no sunlight reaches the surface for more than a year, such that photosynthesis is impossible and continents and oceans cool by as much as 28 degrees Celsius and 11 degrees Celsius, respectively. The absorption of light by the soot heats the upper atmosphere by hundreds of degrees. These high temperatures, together with a massive injection of water, which is a source of odd-hydrogen radicals, destroy the stratospheric ozone layer, such that Earth's surface receives high doses of UV radiation for about a year once the soot clears, five years after the impact.

Temperatures remain above freezing in the oceans, coastal areas and parts of the tropics, but photosynthesis is severely inhibited for the first 1-y [year] to 2-y, and freezing temperatures persist at middle latitudes for 3-y to 4-y. Refugia [areas in which a population of organisms can survive a period of unfavourable conditions.[2]] from these effects would have been very limited. The transient climate perturbation ends abruptly as the stratosphere cools and becomes supersaturated, causing rapid dehydration that removes all remaining soot via wet deposition.[3]

Not everyone agreed with the 'global fires' hypothesis. In March 2009, for example, Claire M. Belcher *et al.* stated that 'non-marine K-T BIRs (Boundary Impact Rocks) from across North America, contain only rare occurrences of charcoal yet abundant non-charred plant remains' – i.e. revealing that some, if not all of the vegetation had not been burnt.[4]

Furthermore, 'Belcher *et al.* revealed that K-T boundary impact rocks (BIRs) from 8 sites, spanning the length of the western interior of North America, contain significantly less charcoal than the associated Cretaceous and Tertiary rocks. Furthermore, on average 99% of organic matter in the K-T BIRs is not charred.' Additionally, they found:

> Comparisons of previously described soot from marine K-T BIRs with modern soot from wildfires suggests that its morphology is

inconsistent with a biomass source, but that it is characteristic of soots produced from partial combustion of hydrocarbons. Moreover, it has recently been revealed that both marine and non-marine K-T BIRs contain carbon cenospheres, which are known only to be formed by combustion of hydrocarbon material. [Cenosphere: a porous or hollow carbonaceous sphere-like particle formed during pyrolysis. Pyrolysis: decomposition brought about by high temperatures.[5]]

The team also argued that even though an abundance of pyrosynthetic polycyclic aromatic hydrocarbons (pPAHs) were in marine KT boundary impact rocks (BIRs), and 'Wildfire is the dominant source of pPAHs in the fossil record throughout the majority of geological time', nevertheless, the K-T impact vapourised a huge volume of rock that is known to have contained hydrocarbons and a significant fraction of organic material, making these a potential source of pPAHs found at the K-T boundary'. The team also noted that 'the reflectance of the K-T charcoals is moderate and certainly does not indicate extremely high temperatures of formation', as for example would have been produced by global fires. (Reflectance: a property of a surface equal to the proportion of incident light that it reflects or scatters.[6])

The team concluded as follows: 'Soot found in the K-T BIRs shows a morphology consistent with soots produced from the combustion of hydrocarbons'. (Hydrocarbon: a compound of hydrogen and carbon, such as any of those that are the chief components of petroleum and natural gas.[7])

Moreover, both marine and non-marine K-T BIRs contain carbon cenospheres. Carbon cenospheres are thought to derive solely from incomplete combustion of pulverised coal or fuel-oil droplets, which suggests that the impact may have combusted an organic-rich target crust. Harvey *et al.* [Mark C. Harvey, Department of Geological Sciences, Indiana University, Bloomington, Indiana, USA] highlight that the Chicxulub impact crater is adjacent to the Cantarell oil reservoir, one of the most productive oil fields on Earth, suggesting that an abundance of organic carbon in the Chicxulub target crust was likely to have been above global mean values.[8]

Even if Chicxulub's organic-rich location is discounted, it has been shown that the global mean crustal abundance for fossil organic matter is more than adequate to account for the observed concentrations of both carbon cenospheres and soot, therefore making the global wildfire hypothesis unnecessary.⁹

Whatever the origin of the KTB soot, the outcome for the Earth's environment and ecology is likely to be the same. Further quantification of what this outcome might have been was provided by Charles G. Bardeen of the National Center for Atmospheric Research, Boulder, Colorado, USA, *et al.*, in their paper entitled 'On Transient Climate Change at the Cretaceous–Palaeogene boundary due to Atmospheric Soot Injections', published in August 2017. By taking Wolbach's measurements for the average amount of soot at all the KTB sites that she examined, and multiplying this by the surface area of the Earth (including the oceans), an approximation for the total quantity of soot in the KTB layer is reached. This is assuming that the soot was distributed globally as seems to be the case.

Said Bardeen, the KTB layer worldwide 'contains as much as 56,000 Tg [55,000 million tons] of elemental carbon, of which 15,000 Tg [14,700 million tons] is in the form of fine soot nanoclusters, and the remaining 41,000 Tg [40,300 million tons] is made up of coarse soot particles'. (Tg: teragram. 1 Tg is 1 trillion grams. Nanoclusters: clusters of nanoparticles. Nanoparticle: a nanoscale particle. Nanoscale: a scale involving dimensions of less than 100 nanometres. Nanometre: 1,000 millionth of a metre.)

The Source of the Soot

Wolbach *et al.* had assumed, said Bardeen, that the observed soot at the KTB 'was caused by global fires ignited by heat from falling spherules'. This soot 'is believed to originate from global wildfires ignited after the impact of a 10 km (6 mile) diameter asteroid on the Yucatán Peninsula, 66 million y [years] ago'. This diameter for the asteroid falls within the range suggested by Luis Alvarez *et al.* of between 3.7 and 8.7 miles. These were not 'normal forest fires but would have acted like large firestorms'.

Bardeen *et al.*'s Experimental Model

In reference to his team's research into the Chicxulub impact, Bardeen stated:

> In this study, we present simulations of the short-term climate effects of massive injections of soot into the atmosphere following the impact of a 10 km [6 mile] diameter asteroid. We assume that the soot

originated from global or near-global [i.e. not covering the entire landmass of the globe] fires.

Given the range of estimates for the fine soot produced by the impact, we consider [hypothetical] soot injections [i.e. soot hurled into the atmosphere as a result of the asteroid impact] of 15,000 Tg and 35,000 Tg.

However, said Bardeen, whichever of the two figures was chosen for the study made little difference, since 'the climate impact of either 15,000 Tg or 35,000 Tg soot injections is similar, because the soot burden and atmospheric residence time [i.e. the time the soot remained in the atmosphere] are sufficiently large in both instances to produce severe, multi-year reductions in surface solar flux'. (Flux: the amount of radiation or particles incident on an area in a given time, in this case referring to sunlight.[10])

Soot as the Dominant Factor

Said Bardeen, 'The short-term climate effects of the soot would augment and probably dominate those of other materials injected by the impact.'

The Sequence of Events

Soot is assumed to be produced by global fires ignited as debris from the impact falls through the atmosphere at high velocities and heats up to very high temperatures. We assume that fine soot is lofted to the upper troposphere in pyrocumuli. Both fine and coarse soot are injected over 24 h [hours]. (Pyrocumuli: dense, cumuliform clouds associated with fire. Cumuliform: from cumulus: cloud, forming rounded masses heaped on each other above a flat base at fairly low altitude.[11])

As regards the coarse soot particles, however, they 'are more likely to have remained lower in the atmosphere', and would have been 'removed rapidly' (presumably by rainfall). Therefore, coarse soot 'plays a negligible role in forcing climate change'. (Climate forcing: the difference between insolation absorbed by the Earth and energy radiated back to space. Insolation: the amount of solar radiation reaching a given area.[12]) In their 'various simulations', the researchers accordingly focused on fine soot in particular.

In summary, Bardeen stated:

Following injection into the atmosphere, the soot is heated by sunlight and lofted to great heights, resulting in a worldwide soot aerosol layer that lasts several years. As a result, little or no sunlight reaches the [Earth's] surface for over a year, such that photosynthesis is impossible, and continents and oceans cool by as much as 28°C and 11°C, respectively.

The ocean does not cool as much as the land 'due to its higher thermal inertia'.

The Ozone Layer

The absorption of light by the soot heats the upper atmosphere by hundreds of degrees. These high temperatures, together with a massive injection of water [presumably the result of burning vegetation – see below, and also of the combustion of oil in the Chicxulub oilfields in the vicinity of the asteroid impact], which is a source of odd-hydrogen radicals, destroy the stratospheric ozone layer. (Odd hydrogen: a chemical family, comprising hydrogen atoms [H], hydroxyl radicals [OH] and hydroperoxyl radicals [HO_2].)

The outcome is that five years later, once the soot has cleared, 'Earth's surface receives high doses of UV radiation for about a year.'[13]

Sulphur

At Chicxulub, the target rock with which the asteroid impacted includes a 2-mile-thick sequence of carbonites (mainly sedimentary limestone) and evaporates. (Evaporite: a natural salt or mineral deposit left behind after the evaporation of a body of water.[14]) The evaporites include gypsum and anhydrites, which contain large concentrations of sulphur.[15] (Gypsum: a soft white or grey mineral consisting of hydrated calcium sulphate. Anhydride: a compound obtained by removing the elements of water from a particular acid.[16])

Bardeen *et al.* estimated that the total quantity of sulphur injected into the atmosphere as a result of the Chicxulub impact was about 100,000 Tg (98,400 million tons).[17]

Elisabetta Pierazzo of the Planetary Science Institute, Tuscon, Arizona, USA, *et al.* used 'atmospheric models coupled to a sulfate aerosol model to investigate climate-forcing and short-term response to stratospheric sulfate aerosols produced by the reaction of S [sulphur]-bearing gases and water vapor released in the Chicxulub impact event'. (Sulphate aerosol: tiny droplets of a solution of a sulphate. A sulphate molecule consists of one atom of sulphur and four atoms of oxygen.)

A sulphate aerosol can absorb sunlight, and also act as a reflector, scattering light that would otherwise have fallen on the Earth.

Pierazzo concluded that 'the introduction of large amounts of sulfate aerosols in the stratosphere results in a significant cooling of the Earth's surface'. However, because 'the geologic record shows no sign of a significant climatic shift across the KT boundary', this was 'indicative of a fast post-impact climatic recovery'.[18]

Soot-Induced Darkness

Bardeen said: '... thus, both soot and sulfate aerosols are sufficient to produce large, transient decreases in global temperature, but large injections of soot will also cause near-total darkness at the surface for a protracted period'.

The Alleged Global Fires

If the 'larger estimates of the mass of soot in the K–Pg boundary layer' were adopted for the study (i.e. 35,000 Tg of soot), said Bardeen, this quantity of soot would 'require [i.e. could only be accounted for by the] burning of all of the above ground biomass [organic matter]'. Wolbach had come to the same conclusion, as already mentioned. In other words, accounted for by fire consuming the entire quantity of vegetation on the Earth's surface: a truly catastrophic prospect for life on Earth! 'Following the impact', Bardeen continued, 'most land experiences desert conditions, and almost no land is capable of supporting plants.'

The Consequences

Any large creatures that managed to survive the global fires may have had trouble locating food in a burned-over landscape. Small creatures adapted to living below the surface or in wet environments, where they could escape fires, and that consumed small amounts of food, are the observed survivors of the K–Pg impact. In addition, some animals can survive for long periods without food and often do not feed during hibernation. (Hibernation: To hibernate is to spend the winter in a dormant state.[19] The word derives from the Latin *hiberna* ['winter quarters'], and in this context it is of great significance, as will shortly be seen. Dormant: having normal physical functions suspended or slowed down for a period; in or as if in a deep sleep.[20])

'Some organisms have dormant stages (for example, in response to prolonged darkness in the polar night) and might be able to survive a long period of darkness with no food.' For example, 'It has been suggested, based on body mass, that large marine ectotherms [such as mosasaurs and plesiosaurs] could have survived starvation for 1,000 d [days].' In such cases, the creature's minimal energy requirements are met by metabolism of fat from their fat stores.

Aftermath

Said Bardeen:

> Placing 15,000 Tg of fine soot into our global climate model shows that 95% of the soot would be removed from the atmosphere in a year, defining the timescale that is represented by the layer in land deposits. On the other hand, it might take decades for the small soot particles to fall to the bottom of the oceans, assuming no zooplankton were present to consume it and excrete it in large fecal pellets.
>
> Temperatures remain above freezing in the oceans, coastal areas, and parts of the Tropics, but photosynthesis is severely inhibited for the first 1 y [year] to 2 y, and freezing temperatures persist at middle latitudes for 3 y to 4 y. Refugia [areas in which a population of organisms can survive a period of unfavourable conditions[21]] from these effects would have been very limited.

Finally, as the stratosphere cools, it becomes supersaturated with water and the resultant precipitation [rainfall] 'removes all remaining soot via wet deposition'.[22]

Conclusion

As already mentioned, it is likely that the Chicxulub and Shiva asteroid impacts (and perhaps others as yet unidentified) occurred virtually simultaneously. The above studies show that the Chicxulub impact alone had drastic global consequences, on both land and in the sea. The food chain collapses from the bottom up; in a knock-on effect, the larger herbivorous terrestrial and marine animals died of starvation, whereupon the larger carnivores that prey upon them are likewise deprived of their food supply and also perish. The larger creatures are disproportionately impacted, simply because their food requirements are greater.

Questions that Remain

Wolbach's notion that the Chicxulub asteroid impact caused the entire Earth's biomass to ignite and burn is somewhat counterintuitive, disruptive as the event was. Others may prefer Belcher *et al.*'s proposition that the presence and quantity of KTB soot resulted from the vapourisation of local hydrocarbon-containing target rocks at the impact site.

Nonetheless, the principal mechanism for the extinction of a large proportion of the Earth's flora and fauna was a failure of photosynthesis, caused by both soot and sulphate aerosols in the atmosphere, thus drastically reducing the amount of sunlight falling on the Earth. Other factors were acid rain (produced in the atmosphere when sulphur dioxide dissolves in rain water) and contamination caused by soot and other ejecta products.

Dinosaurs *versus* the other Reptiles and the Mammals

Clearly, the Chicxulub and Shiva impacts were a colossal insult to the entire ecosystem of the Earth, and the authors cited above have produced compelling evidence for an asteroid-induced global failure of photosynthesis on both land and in the sea. However, this does not explain why the dinosaurs became totally extinct, while other species of reptile and also of mammals (some of which were as large, if not larger, than the smaller dinosaurs) survived when faced with exactly the same conditions as led to the global KT extinction.

Chapter 20

Some Terrestrial, Semi-Aquatic and Marine Creatures that Survived the KT Extinction Event

Arthropods

As already mentioned, arthropods are cold-blooded invertebrate animals such as insects, spiders or crustaceans.

Bumblebees, for example, are flying insects that had evolved by about 100 mya (Mid-Cretaceous). 'The United States Geological Survey estimates there are currently 20,000 species of bees' in existence today.[1]

The queen bee is the single reproductive female in a colony of honeybees.[2] The virgin queen bee flies on a warm, sunny day to mate in flight with a dozen or so drones (male bees). She stores their sperm in her body and releases it during the rest of her life – which might be up to seven years.

The worker bees dig down 4 inches or more into well-drained soil and excavate a small hole in which they will spend the winter, surviving temperatures down to −19 degrees Celsius. However, only the queen bumblebee hibernates during the winter months. In spring, she emerges to lay her eggs. So here is the word 'hibernation' cropping up again.

For the queen bee, hibernation can last for as long as nine months. Entering hibernation is important because it protects the queens from the risks of predation, starvation, disease, and adverse weather conditions above ground.[3]

Mammals

The three main groups of living mammals are the monotremes, the placentals and the marsupials.

(1) Monotremes

These are mammals of the small order Monotremata that are noted for laying eggs (unlike other mammals) and having a common urogenital and digestive opening.[4]

There are only five living species of monotreme extant today: the duck-billed platypus, and four species of echidna (spiny anteater). But the echidna evolved only as recently as 20 mya to 50 mya (i.e. post-KT event), having descended from a platypus-like monotreme. Monotremes are found only in Australia and New Guinea.

In 2008, the fossilised jaw bone of a platypus (*Teinolophos trusleri*) was discovered in Australia and dated to between 121 and 112.5 mya (Early Cretaceous). The platypus is semi-aquatic, inhabiting small streams and rivers in Tasmania and Queensland, Australia. It is a carnivore that feeds on annelid worms, insect larvae, freshwater shrimp and freshwater yabby, which it digs out of the river bed with its snout or catches while swimming. (Annelid: segmented worms of the phylum Annelida, which includes earthworms, lugworms, ragworms and leeches. Yabby: a genus of small aquatic crayfish.[5])

The platypus produces milk that oozes from the skin of its belly. This enables its young to develop without having to leave their mother's burrow, which is usually built into the bank of a river bed.

Critic, novelist and poet G.K. Chesterton (1874–1936), in his poem 'The Donkey', described that animal because of its shape as 'The devil's walking parody/On all four-footed things'. In the world of the monotremes, the same could be said of the platypus, with its elongated snout (hence the term 'duck-billed platypus'), broad, flat tail, and webbed feet.

How was the platypus able to survive the KT extinction event when the evidence is that it is not known to hibernate, although it can enter a state of torpor for up to six days? (Torpor: a state of mental or physical inactivity; lethargy.[6])

Perhaps hibernation was a survival mechanism that it possessed at that time but has since lost. Alternatively, being a small carnivore at home both on land and in the water, this would have given it an advantage survival-wise. But the truth is that how it survived the KT extinction remains a mystery.

(2) Placentals

Placentals probably evolved about 90 mya (Late Cretaceous). (Placental: denoting mammals that possess a placenta.[7] Placenta: an organ within the uterus of pregnant eutherian mammals, which nourishes and maintains the foetus through the umbilical cord. Eutherian: mammals of the major group Eutheria, which comprises the placentals.[8]) The foetus is thereby safely protected within the mother's womb.

The common tenrec (*Tenrec ecaudatus*), for example, is a small, insectivorous, placental mammal found in Madagascar and the Comoro Islands, which had evolved by the Late Cretaceous. A hedgehog-like creature, it weighs a mere 2 pounds.

In 2014, Barry G. Lovegrove, evolutionary physiologist, University of KwaZulu-Natal, South Africa, *et al.*, published a report, entitled 'Mammal Survival at the Cretaceous-Palaeogene boundary'. When radio transmitter devices were fitted to fifteen tenrecs, said Lovegrove, it was discovered that 'one adult male hibernated for 9 months until we were forced to dig it up, because the radio transmitter batteries were dying'. This was 'a good indication that tenrec ancestors were also heavy sleepers; capable of hibernating through something as catastrophic as the event that wiped out the dinosaurs'.[9]

Said journalist Jenna Iacurci, who reported on the study for *Nature World News*:

> The tenrec was found to hibernate for long stretches without waking, for as long as 9 months straight. Males apparently, start hibernating between February and April. Whereas, females commence as late as May. Both sexes were reported to arouse in October and November which is the mating season.[10]

(3) Marsupials

These are members of the order Marsupialia. Marsupial mammals first evolved about 75 mya (Late Cretaceous). The young of Marsupialia are born incompletely developed and are carried and suckled in a pouch on the mother's belly.[11] Within the pouch are teats through which milk is channelled. This has obvious evolutionary advantages in that the pouch provides protection for the young. Furthermore, each pouch can hold more than one infant at a time.

The opossum (or 'possum') is a marsupial of the infraclass Marsupialia, order Didelphimorphia, family Didelphidae, endemic to the Americas.

In 2009, Inés Horovitz *et al.* of the Division of Paleontology, American Museum of Natural History, New York City, USA, announced the results of their study of the origin of opossums. The team concluded that 'the basal splitting of Marsupialia into Didelphimorphia and all the other marsupials' had already occurred by 'at least about 65.18 mya' (i.e. immediately after the KT Event, in the Danian age of the Paleocene Epoch, that immediately succeeded the Cretaceous). It is, therefore, highly likely that opossums of some description were alive at the time of the KT extinction event.[12]

In an article published in *Australian Geographic* in July 2016, entitled 'Switching off: Hibernation in Australia', Shannon Verhagen stated:

> In Australia four species of pigmy possum are known to hibernate for extended periods of the year. The Eastern Pygmy Possum, found along the east coast from Queensland to South Australia and throughout Tasmania, holds the title for the longest hibernation of any mammal in the world, and is capable of spending a whole year in the less active state.[13]

Reptiles

What is called 'hibernation' in mammals is called 'brumation' in reptiles. The two processes are very similar, but differ in the following respects. During brumation, reptiles use glycogen in their muscles instead of fat reserves; reptiles are inactive for long periods, but this is punctuated by bouts of activity. For example, they crawl out of their burrows when it is warm outside and bask in the sunshine. Mammals neither eat nor drink during hibernation, whereas during brumation, reptiles need to drink water to avoid dehydration.

Crocodilians, for example, had evolved by 95 mya (Mid-Cretaceous). Said Casey Holliday of the University of Missouri: 'The Cretaceous [Period] is full of giant crocs including [of the genus] *Sarcosuchus*, *Dyrosaurus*, *Deinosuchus*, "Shieldcroc" [*Aegisuchus*], and others. Today, 23 different crocodilian species exist.'[14]

The crocodile is a large aquatic reptile to be found throughout the tropics in Africa, Asia, the Americas and Australia.

Said science writer Alina Bradford: 'Crocodiles tend to congregate in freshwater habitats like rivers, lakes, wetlands, and sometimes in brackish water or even in some saltwater regions'. Crocodiles are carnivores, which means they eat only meat. In the wild, they feast on fish, birds, frogs and crustaceans (they are also opportunistic devourers of terrestrial creatures such as large mammals). Crocodiles 'are cold-blooded and cannot generate their own heat'.[15]

As winter approaches, crocodiles dig out burrows in the banks of rivers, lakes or ponds to brumate for up to eight months. In fact, a crocodile is believed to be able to go without food for up to two years.

Reptiles do not normally feed during brumation, but a moist environment such as those described above is essential to keep them from becoming dehydrated.

Amphibians

An amphibian is defined as a cold-blooded vertebrate animal of the class Amphibia that comprises frogs, toads, newts, salamanders, and burrowing, worm-like caecilians.[16] In adult form, amphibians possess lungs.

The oldest frog fossil is of a species called *Triadobatrachus massinoti*. About 3.9 inches long, it dates from about 250 mya (Early Triassic).

Terrestrial frogs begin life in an aquatic environment as tadpoles, before migrating to live on land. In winter, they typically burrow down below frost level and create burrows, called 'hibernacula' (hibernation space), or they take refuge in cavities or crevices. In extreme cold, they may stop breathing, their hearts may stop beating, and they may actually freeze solid. However, high concentrations of glucose within their bodies prevent the formation of ice crystals, which would otherwise kill them.

Aquatic frogs live mainly in water shallow enough for them to rise to the surface to breathe. In wintertime, they typically hibernate underwater, absorbing oxygen through their skins.

Fish

An examination of fossilised scales indicates that the earliest known sharks date from at least 420 mya (Late Silurian Period of the Palaeozoic Era). An examination of sharks' teeth indicates that with the KT extinction, the largest sharks vanished, and those species that survived were far smaller and less varied than they had been before.

The Greenland shark (*Somniosus microcephalus*) is native to the north Atlantic waters around Greenland. It may grow to a length of 20 feet and weigh 2,500 pounds.

The Greenland shark is able to tolerate water temperatures down to −1 degrees Celsius. As the temperature drops, the shark becomes increasingly inactive as its metabolism slows. However, it continues to swim, albeit slowly. Its condition is therefore one of torpor, rather than a true hibernation.

These creatures prefer cold water, and normally live at depths of up to 7,000 feet. But in winter, they migrate nearer to the surface, where the water is colder than that on the seabed.

The bodies of Greenland sharks 'contain high levels of trimethylamine oxide (TMAO), which helps to regulate their osmotic pressure'.[17] (Osmoregulation: the maintenance of constant osmotic pressure in the fluids of an organism by the control of water and salt concentrations.[18])

The TMAO, together with unusually high amounts of urea, also acts as an antifreeze.

Having the ability to enter a state of torpor and being able to withstand cold in the manner described would have given sharks an immense edge when it came to surviving the KT extinction.

Conclusion

The creatures described above coexisted with the dinosaurs but survived the KT extinction event (as outlined above), whereas the dinosaurs perished entirely. So what was it in the constitution of the survivors – something that the dinosaurs lacked – that made this possible? Was it related to their metabolic make-up, their ability to hibernate, brumate, or enter a prolonged state of dormancy/torpor? This will be discussed in more detail shortly.

Chapter 21

Were Dinosaurs Warm-Blooded?

To address this question, it is important to study how dinosaurs evolved.

Tetrapods are a superclass of animals (Tetrapoda) that first appeared about 367.5 mya in the Late Devonian Period. Tetrapods are four-limbed vertebrates (with few exceptions, such as snakes), which evolved from fish. The class includes reptiles (including dinosaurs and birds), amphibians and mammals – all of which possess a heart and circulation.

Reptiles, from which the dinosaurs evolved, are cold-blooded. Birds, however, which are also classed as dinosaurs, are warm-blooded. The question as to whether dinosaurs were warm or cold-blooded is still debated. However, French palaeontologist Armand de Ricqlès (born 1938) discovered Haversian canals in dinosaur bones and argued that they were evidence of endothermy in dinosaurs. (Haversian canals: minute tubes that form networks in bone and contain blood vessels.[1]) These canals are common in 'warm-blooded' animals, he said.[2]

The likelihood is, therefore, that dinosaurs were indeed endothermic (warm-blooded).

Because a small dinosaur had a relatively high surface area-to-volume ratio, and therefore a higher potential for heat loss compared with a large one, the former would be particularly vulnerable to cold. But as already mentioned, the smaller dinosaurs were comparable in size to the larger mammals, and as mammals did not become extinct, a cold environment could not have been the sole reason for the dinosaurs' extinction.

Chapter 22

The Ability to Hibernate or Brumate: The Critical Factor

The survival of any animal depends on whether it is suitably adapted to survive in the physical environment in which it finds itself, and also on whether it is able to successfully compete with its competitors for food, living space, water, etc.

The fact that the dinosaurs survived on Earth for about 181 million years indicates that they were ideally adapted for life on the planet – that is, until the KT extinction event. Then, unlike many other species of reptile and many species of mammal, because they lacked some particular capability that would have rendered them able to survive the cataclysmic asteroid (or twin asteroid) event, they perished.

What was the predominant factor that led to the extinction of the dinosaurs? Was it the choking, toxic, post-impact atmosphere? Probably not, because this would also have killed the mammals. Was it the intense heat generated by the impact, together with the alleged global fires that consumed the Earth's biomass? Again, probably not, for the same reason. In any event, it is quite likely that some of the dinosaurs would have kept cool by entering caves or plunging into water. Was it the intense cold? Again, probably not, because evidence is emerging that many dinosaurs were covered in fur and feathers, to a greater or lesser extent.

However, what no dinosaur could survive was starvation for any length of time. Following the impact of an asteroid (if not two, or even more), the Earth was covered in a layer of soot, perhaps several feet thick; sunlight was blocked by the smoky ejecta; and the planet became extremely cold. The outcome would be the interruption of photosynthesis, probably for many months, if not years, whereupon the bottom end of the terrestrial (and also the marine) food chain collapsed. Given this likely scenario, as regards the dinosaurs, one may assume that first the herbivores died of starvation, followed by the carnivores that depended on the former as their source of food (to the carnivores, anything smaller would have been regarded as a mere snack). With their extinction, the long reign of the dinosaurs came to an end (as it did for many other species of flora and fauna, as already mentioned).

A closer look at some of the creatures that survived the KT extinction event reveals that they all had a particular capability in common – that is, the ability either to hibernate or, in the case of reptiles, to brumate.

Mingke Pan of the Department of Life Sciences, Imperial College, London, stated as follows:

> Hibernation is an adaptation mainly found in mammals and birds that occurs at physiological, behavioural, and molecular levels to sustain life during winter months when resources are insufficient or unpredictable. It is generally associated with a profound reduction in core body temperature to the level of ambient temperature, and a reduction in metabolic rate to a fraction of basal metabolic rate (BMR), parallelled with bradypnoea [reduced respiratory rate] and reduced heart rate.[1] (Metabolism: the chemical processes that occur within a living organism to maintain life.[2] Heterothermy: a physiological term for animals that vary between self-regulating their body temperature and allowing the surrounding environment to affect it.[3])

In short, hibernation is a survival mechanism that enables a creature to survive the winter months. By entering a state of hibernation, said Mingke Pan, the advantage is that 'hibernators can produce a remarkable saving in energy consumption. [For example] In winter, yellow-bellied marmots are able to save up to 85% of their energy, which will greatly enhance winter survival.' Pan drew a distinction between hibernation and torpor:

> Similar to hibernation, torpor is a survival tactic involving a reduction in body temperature and metabolic rate. The physiological definitions of torpor and hibernation remain unclear, but they are used in the scientific literature to express different durations and amplitudes of hypometabolic state. Hibernation generally has a longer duration of up to 9 months; whereas torpor lasts < 24 h [hours] and is commonly referred to as daily torpor. [In fact, torpor may in certain species last for longer periods, as already indicated.] The mean minimum metabolic rate was observed to reach 6% of BMR during hibernation but 35% of BMR during torpor.[4]

Is the Ability to Hibernate or Brumate an Ancient Trait?

Said Gordon C. Grigg of the School of Biological Sciences, University of Queensland, Brisbane Australia: 'The most compelling evidence that hibernation is plesiomorphic [i.e. a characteristic that is shared with and derived from an ancestral clade] is that it occurs with essentially similar patterns in the three extant mammalian subclasses.' These subclasses are Prototheria (comprising the monotremes); Metatheria (comprising the marsupials); and Eutheria (comprising mammals that possess a placenta). The genetic aspects of hibernation will be discussed shortly.

Grigg referred to the 'spectacular capacities for hibernation' seen in such modern-day mammals as Arctic ground squirrels, mountain pygmy possums and short-beaked echidnas. This 'implies that these capabilities are specializations from attributes which can be traced back to the first protoendotherms [precursors of the endotherms]'.[5] This begs the question, when did mammals first become warm-blooded (endothermic)? The answer appears to be around 259–251.9 mya (during the Late Permian Period; see below).

Not only that, said Grigg, 'some of the (physiological and behavioural) attributes that support hibernation in mammals' can be traced back even further, 'to the reptiles'.

Grigg continued: 'Hibernation was assumed by many authors to be an advanced character [i.e. one of comparatively recent origin]; evolved to allow mammals to cope with living in cold environments. It was regarded as an apomorphic capability.' (Apomorphic: distinguishing an organism or taxon from others that share the same ancestor: i.e. non-ancestral.)

However, 'Since the recognition that hibernation occurs not only in eutherians but in a couple of extant marsupials and one of the three extant monotremes (the echidna) as well, it has become accepted by most people that hibernation is plesiomorphic [i.e. a character derived from an ancestral clade] in mammals.'

Also, it is now realised that hibernation (and torpor) occurs not only in response to cold but also in response to dry conditions, and that many of the biochemical and physiological capacities that characterise it were in the reptiles, many of which, of course, also abandon their normal behaviour for periods of inactivity, to go torpid during a cool or dry season. So, hibernation (and endothermy) can be seen as elaborations of a pre-existing 'chassis' [basic framework]. Think of it as a continuum.[6]

Said Kevin Rey of the Evolutionary Studies Institute, University of Witwatersrand, Johannesburg, South Africa: '... by comparing the ratios of oxygen isotope in fossils we were able to show that a group called the Cynodontia acquired warm-bloodedness somewhere during the Late

Permian Period [i.e. about 259–251.9 mya]'.[7] (Cynodonts or primitive synapsids: reptiles of the Late Permian and Triassic Periods, the members of which show increasingly mammalian characteristics and include the ancestors of mammals.[8])

In other words, it is likely that the ability which some mammals have to hibernate has existed ever since their reptilian ancestors (the cynodonts) became warm-blooded (endothermic) more than 251 mya. Furthermore, many mammals retained the ability to hibernate, whereas the reptiles evidently lost this ability and instead evolved the capacity to brumate, and this explains why certain mammals and certain reptiles were able to survive the KT extinction event.

Mammalian Evolution

The first vertebrates were fish that lived about 500 mya (Cambrian Period of the Early Palaeozoic Era). The first amphibians developed from fish about 370 mya (Devonian Period of Late Palaeozoic Era). Amphibians were the first terrestrial vertebrates.

Amniotes are a clade of terrestrial tetrapod (having four legs and feet) vertebrates comprising reptiles, birds and mammals. The early amniotes, which had the appearance of small lizards, evolved some 312 mya (Late Carboniferous Period). Amniotes lay their eggs on land or retain the fertilised egg within the mother (unlike the anamniotes – fishes and amphibians that typically lay their eggs in water).[9]

The amniotes evolved into two main lines: the Sauropsids and the Synapsids. Sauropsids are a group of amniotes that includes all existing and extinct reptiles and birds. Synapsids are a group of amniotes that includes all existing and extinct mammals. The earliest mammals appeared between 235 and 201 mya (Late Triassic).[10]

How do Creatures Prepare for Hibernation?

Some preparation is required of the would-be hibernator, said author and freelance writer Ed Grabianowski:

> Some animals prepare a den, and line it with insulating material, such as leaves or mud. Food can be kept in the den if it is non-perishable, but this requires the animal to wake up briefly during the winter to eat. Another option is to eat a large amount of food starting in late summer and building up a reserve of internal fat. Some animals even do both.[11]

What Triggers Hibernation?

The trigger for hibernation, said Grabianowski, was the temperature of the creature's environment:

> When it gets cold outside, animals get ready to hibernate. When it warms up, they wake up. Therefore, hibernation periods can vary depending on the weather that year. An Indian summer [a period of dry, warm weather occurring in late autumn[12]] and an early thaw could result in a very short hibernation.[13]

The Physiology of Hibernation

In regard to the animal's metabolism, said Grabianowski: 'Hibernation is mainly controlled by the endocrine system.'[14]

Endocrine is defined as of or denoting glands that secrete hormones or other products directly into the blood or lymph.[15] It is the release of hormones by the pituitary gland that, in turn, regulates the release of hormones by the other endocrine glands. They include:

- The thyroid: Secretes thyroxine. This regulates metabolism.
- The pancreas: Endocrine cells in regions of the pancreas called the islets of Langerhans secrete the hormone insulin. This regulates blood glucose.
- Adrenal glands: Secrete adrenaline, noradrenaline and corticosteroids. Adrenaline regulates rates of blood circulation, respiration and carbohydrate metabolism. Noradrenaline functions as a neurotransmitter and regulates blood pressure. Corticosteroids regulate stress response, immune response, inflammatory response, carbohydrate and protein metabolism, blood electrolyte levels, etc.
- Pineal gland: Situated in the epithalamus of the brain, it secretes the hormone melatonin that regulates the growth of an animal's winter coat.

Continued Grabianowski:

> When a mammal enters hibernation, it becomes somewhat like a cold-blooded animal. Its body temperature will vary, depending on the temperature around it. However, there is minimum known as a 'set point'. When the mammal's body temperature reaches this set point, the metabolism kicks in and burns some fat reserves. This generates some energy, which is in turn used to heat things back up

above the set point. Larger animals have a higher set point. If they let their temperature drop too low, it would require an enormous amount of energy to heat back up again.[16]

This process is regulated by genes. Mingke Pan stated: 'During natural hibernation, there seems to be a switch from a glucose metabolism to a lipid metabolism as genes involved in glycolysis are under-expressed while genes for fatty acid catabolism are over-expressed.'[17] (Glycolysis: the breakdown of glucose by enzymes, releasing energy. Catabolism: the breakdown of complex molecules in living organisms to form simpler ones, together with the release of energy.[18])

Grabianowski went on to describe how the process of thermogenesis works (Thermogenesis: the production of bodily heat.[15]):

Hibernating animals have a special way to stay warm: brown fat. Regular body fat is white. When it's burnt, it's broken down through intermediate steps to fuel shivering, which in turn produces body heat as the muscles are exerted. Brown fat, however, has a special protein that allows it to skip the middleman. Brown fat can be oxidised directly by cell mitochondria (the 'power stations' of cells). This chemical reaction produces heat directly, without intermediate steps and without [the requirement for] shivering. The scientific term for the process is 'non-shivering thermogenesis'.[19]

Excrement

The animal that is hibernating is not eating but burning fat. Therefore, said Grabianowski, 'no fecal matter is produced because nothing is passing through the digestive tract and intestines'. But the body continues to produce urea, 'the waste product that is the main component of urine'. This would normally be a problem. However, bodies of hibernating animals 'are able to recycle the urea'. In this way, they avoid dehydration: 'They are also able to extract enough water from their own body fat to stay hydrated.'[20]

Is it certain that Dinosaurs did not Hibernate?

In 2005, the fossilised skeletal remains of an adult and two juvenile *Oryctodromeus cubicularis* dinosaurs were discovered within a burrow, in the Mid-Cretaceous Blackleaf Formation of southwest Montana, USA.[21] These were small, bipedal, of the order ornithischia, and about the size of a modern-day coyote.

From this single discovery, may one conclude that this and possibly other species of dinosaurs hibernated? The answer is no, for the following reasons. Firstly the burrow may have been created by another creature, and the dinosaur may have been using it opportunistically to give birth to its young in safety. Secondly just because an animal is a burrower/den maker, does not necessarily make it a hibernator. For example, in the modern world, mammals such as rabbits, badgers, and foxes fall into this category.

Holly Woodward of the Department of Earth Sciences, Montana State University, USA, analysed the fossilised bone tissue of seventeen dinosaurs, each about the size of the average dog, discovered in Australia's State of Victoria. These dinosaurs had lived between 112 and 100 mya (Early Cretaceous) when this region of Australia lay within the Antarctic Circle and was, therefore, extremely cold.

Research performed on animals alive today indicates that in their skeletal bones, lines of arrested growth (LAGs) form annually, regardless of latitude or climate. However, during periods of hibernation, bones cease to grow. But Holly Woodward found no difference between LAGs in the bones of the seventeen dinosaurs that she examined and the bones of those living elsewhere, indicating that the Antarctic dinosaurs did not hibernate.[22]

Conclusion

Since publicity about dinosaurs is usually related to the largest and fiercest types, it might be concluded that large dinosaurs were too large to hide away in burrows and see out the KT extinction event. But, as already mentioned, many species of dinosaur were smaller than the larger mammals and the larger reptiles, and therefore the ability to burrow and take refuge in this way was not the critical factor in the dinosaur extinction.

What was crucial was the fact that, unlike many species of mammal and fellow reptile, dinosaurs evidently did not have the physiological ability to hibernate or brumate respectively. Had it been otherwise, some species of dinosaur, at least, would have certainly survived the KT extinction event. However, in the circumstances in which the dinosaurs found themselves, nothing could have saved them.

Chapter 23

Hibernation/Brumation/Torpor: Some of the Complexities Involved

Mammals

Genetics: The 'Hibernation Switch'

Said Cheng Chi Lee, Professor of Biochemistry and Molecular Biology at the University of Texas, for a mammal in a cold environment, 'the body tries everything it can to keep the temperature up, until you reach a point where it no longer can, and then the body will cool'.

The gene Clps codes for colipase, a protein co-enzyme secreted by the pancreas and necessary for the breakdown of fat. Lee discovered that when mice are kept in constant darkness, the gene Clps 'switches on in all tissues': '... when I discovered this really fascinating phenomenon, I was wondering, why would nature design such a thing? And it occurred to me the most likely time you will find an animal in a state of constant darkness is when it goes into hibernation or a long period of torpor.'[1]

Therefore, the sequence of events would appear to be as follows. In mammals, prolonged darkness (and/or prolonged cold) triggers hibernation. In dark conditions, the gene Clps activates colipase. This, in turn, breaks down fat to maintain the energy supply to the mammal's cells. This occurs over a prolonged period, until hibernation ceases. However, many more genes come into play during hibernation, as will be seen.

Calcium Pumps: Calcium Retention in Cells

Calcium pumps are a family of ion transporters found in the cell membrane (boundary) of all animal cells. (Ion: an atom or a molecule with a net electric charge – either positive or negative.[2])

Calcium pumps play a crucial role in keeping the intracellular calcium concentration roughly 10,000 times lower than the extracellular concentration. In other words, calcium pumps reduce the entry of calcium ions into the cell, while enhancing their removal from the cell.

It is because of the action of these pumps, said Xiao-Chen Li of Peking University *et al.*, that 'heart tissues from hibernating mammals, such as ground squirrels, are able to endure hypothermia, hypoxia [deficiency in the amount of oxygen reaching the tissues[3]], and other extreme insulting factors'. These conditions would otherwise be 'fatal to humans and [other] non-hibernating mammals'.[4]

However, said Shi-Qiang Wang of Peking University *et al.*, while calcium ion homeostasis is key to the successful functioning of the cardiovascular system, 'regulation in other types of cells, such as neurons and endocrine cells, is also important for mammals to survive hibernation'.[5] (Homeostasis: the maintenance of a stable equilibrium, especially through physiological processes.[6])

Insects

Following the KT extinction event, said Conrad Labandeira of the Museum of Natural History, Smithsonian Institute, Washington, DC, *et al.*: 'Several quantitative studies at the family level indicate that the diversity of insects suffered no decrease beyond normal levels of background extinction.' (Family: a principal taxonomic character ranking above 'genus' and below 'order'.[7])

However, 'below the family level, the demise of the Late Cretaceous aphidioid beetle and genera provides some evidence for extinctions'.[8] (Aphid: a small bug that feeds by sucking sap from plants, especially a blackfly or greenfly.[9])

In addition to the aforementioned bumblebee, it is likely that many other insects of the Late Cretaceous possessed the ability to hibernate at the time of the KT Extinction, just as they do today. Also, like today's insects, it is likely that some insects of that period possessed the following characteristics, which enabled them to survive the cold post-KT impact conditions: a wax coating, which protected their bodies against ice; the ability to synthesise cryoproteins ('antifreeze'); the ability to synthesise ice-nucleating proteins, which enabled them to regulate the formation of ice within their bodies. (Nucleation: a physical process in which a change of state, in this case water to ice, occurs in a substance around certain focal points, known as nuclei.)

Reptiles

In reptiles, prolonged darkness and/or prolonged cold can trigger brumation, when it is likely that similar physiological processes, as described above for hibernation, occur.

Fish

Fish are cold-blooded (poikilotherms). Their metabolism slows dramatically when the temperature in their environment drops. Anyone who keeps goldfish (members of the carp family) in their garden pond will know that these creatures can survive several months of winter cold and darkness, virtually immobile, and without the need for food or oxygen. They can even survive when ice, several inches thick, sits on the surface of the water. In this condition, lactic acid builds up in their bodies, which they are able to convert anaerobically (in the absence of free oxygen) into ethanol, which then diffuses across their gills into the surrounding water, thereby removing the waste product that would have otherwise been lethal to them.

Arthur 'Art' DeVries, US Professor of Animal Biology, University of Illinois, was the first to describe how proteins in notothenioid fish bind to ice crystals in the blood to prevent the fish from freezing. (Nototheniodia: a suborder of the order Perciformes, which include Antarctic and sub-Antarctic fish.)

DeVries published his results in two scientific papers – one with Donald E. Wohlschlag (Department of Biological Sciences, Stanford University, Stanford, California, USA) entitled 'Freezing Resistance in some Antarctic Fishes' in 1969, and another entitled 'Glycoproteins as Biological Antifreeze Agents in Antarctic Fishes' in 1971.

In 1996 his discovery was described by the National Science Foundation, Alexandria, Virginia, USA, in a paper entitled 'Fish Antifreeze Proteins':

> To survive, Antarctic fishes have developed proteins that act as antifreeze. These antifreeze proteins are a group of unique macromolecules that help some polar and sub-polar, marine bony fishes avoid freezing in their icy habitats. The proteins were discovered by Dr Art DeVries from fish that he collected at McMurdo Station, whilst he was a graduate student at Stanford University in the early 1960s.

Waters of the Southern Ocean are so cold that temperate and tropical fish would freeze if they were placed in this environment. Salt in sea water allows it to remain liquid until about -1.9 degrees Celsius; almost 2 degrees below the freezing temperature of fresh water. The antifreeze proteins, along with normal body salts [i.e. salts that occur naturally within the fish's body], depress the freezing point of blood and body fluids to 2.5 degrees Celsius, slightly below the freezing point of sea water. These proteins bind to and inhibit growth of ice crystals within body fluids, through an absorption-inhibition process. The proteins attach to small ice crystals, stemming their growth. The mechanism that inhibits further growth of the ice crystal remains under study, but apparently Antarctic fish are able to survive with very small ice crystals in their body fluids.[10]

Chapter 24

The Genetics of Hibernation

It is now known that for a creature to enter a state of hibernation, numerous genes are required to come into play. For example, in the case of the thirteen-lined ground squirrel (*Spermophilus tridecemlineatus*), no less than 748 'hibernator genes' were identified.

The question arises: does a creature have the capacity to hibernate because it possesses certain genes that the non-hibernator does not? Surprisingly, this is not the case; for, as Mingke Pan pointed out, 'most of the genes associated with hibernation are present in the human genome'. The same also, presumably, applies in the case of reptiles, including the extinct dinosaurs. In other words, the deciding factor in determining whether a creature has the capacity to hibernate is dependent on whether it has the capacity to 'switch on' certain requisite genes, and having been activated, these may be designated as 'hibernator-enabling genes'.

Therefore, although humans, and presumably all other mammals, possess genes that, under certain conditions, could acquire the status of hibernator-enabling genes, it is a fact that all human beings, and many mammals too, are physiologically incapable of hibernating. The assumption must be that for them this capability has been lost over time. However, some mammals did retain the capacity to hibernate and were, therefore, able to survive the KT extinction event.

What of the Dinosaurs?

Dinosaurs are classified as reptiles, so why did they not have the ability to brumate? Probably because, just as human beings possess 'hibernator genes' but have lost the capacity over time to switch them on, so the same applies to the dinosaurs in respect of 'brumator genes', and it was for this reason that every single species of dinosaur perished in the KT extinction event.

By contrast, some species of reptile retained the capacity to activate their brumator-enabling genes, and by virtue of this fact, they were able to survive the KT extinction event.

When did Living Creatures First Acquire the Capacity to Hibernate/Brumate?

Hibernator-enabling genes and brumator-enabling genes are probably similar, if not identical. As already mentioned, some amphibians, which evolved from the reptilian line 370 mya, possessed the ability to hibernate. This suggests that these amphibians had inherited this capability from their primitive reptilian ancestors in the Devonian Period of the Late Palaeozoic Era.

Chapter 25

For the Mammals Shall Inherit the Earth!

To survive the catastrophic KT extinction event, it was necessary, as a general rule, for creatures to possess the ability to hibernate or brumate, and avoid the physical hazards of global forest fires; falling superheated molten spherules; acid rain; soot; and other ejecta. Thus, it is no coincidence that the majority, if not all, of the mammalian and reptilian survivors hibernated or brumated respectively, either below ground in burrows – or at the bottom of ponds, as was the case with aquatic frogs.

That the mammals have taken over from the dinosaurs as the dominant species on the planet is not in dispute. But why were the mammals so successful? Firstly their competitors, the dinosaurs, had been completely eliminated as a result of the KT extinction event. But then there were the reptiles to contend with. However, the mammals had considerable advantages over their reptilian cousins, as will shortly be seen. But to begin at the beginning.

At the Lufeng Basin in the province of Yunnan in southwestern China, palaeontologist Wang Tao, Director of the Lufeng Dinosaur National Geopark, discovered a small fossil around 2cm long. It was the head of a tiny animal, which was given the name *Hadrocodium*. From its teeth, which were the shape of an insect-eating mammal, it was clear that this was no reptile. In fact, it was a fossil of one of the earliest known mammals.

The synapsid line (mammalian) became distinct from the sauropsid line (reptilian) between 325 and 315 mya (Carboniferous Period).

Hadrocodium dates from about 195 mya (Sinemurian Age of Early Jurassic). Today, there are a total of 6,495 recognised living species of mammal, and mammals inhabit every corner of the planet.[1]

Mammals evolved the following features, which enabled them not only to survive but also to thrive and become the dominant species on Earth:

- Warm-bloodedness: this characteristic enabled them to operate even in cold conditions; in particular, at night, when they could hunt (not so the reptiles, which were cold-blooded).

99

- Bodies covered in hair: this provided insulation, protection and enhanced sensation (see below).
- A relatively large brain (see below).
- A sophisticated sense of smell (see below).
- An ability to hear high-pitched sounds (see below).

Using a micro-CT (computerised tomography) scanner, Professor Zhe-Xi Luo of the Carnegie Museum of Natural History and Nanjing University discovered – in respect of the skull of *Hadrocodium* – that the brain was much larger than that which exists in the brain of a reptile of similar size. Furthermore, it contained proportionately large olfactory bulbs (that part of the brain which detects smell, an ability that was especially valuable to these creatures at night-time). Also, a three-bone middle ear enabled them to hear higher-pitched frequencies, which the reptiles could not detect. Again, this was invaluable – for example, when it came to hunting insects in the dark.

A large area of *Hadrocodium*'s brain was concerned with sensation and with the processing of sensory impulses arriving via the nerve receptors at the base of each hair. This enabled the creature to map the world around itself.[2] Altogether, characteristics such as these enabled the mammals to survive, even in the shadow of the dinosaurs, and to thrive in their absence.

Chapter 26

A New Dinosaur is Discovered

They say a book is never finished and with regard to books on dinosaurs, this is definitely true: the reason being that new species are continually being discovered.

For me, I was watching the BBC South Today television news programme on Wednesday, 12 August 2020, when the news broke that a dinosaur, hitherto unknown to science, had been discovered on the Isle of Wight. On most mornings my wife Rachel and I walk along the clifftop at nearby Canford Cliffs in our hometown of Poole in Dorset, from where the island is visible 15 miles to the southeast, on a clear day.

There on the screen was the discoverer, local man and aircraft engineer Paul Farrell, from Ryde on the Isle of Wight, who, while his daughter, Izzy, was having her dancing lesson, found himself with an hour to spare on a lovely evening. Said he, 'I was walking along the beach, kicking stones and came across an odd-shaped fossil.' This was Shanklin beach, near Horseshoe Ledge. 'I picked it up, and nearly threw it back. But I noticed that it had rounded edges, and although I am not a fossil hunter, I realised that it looked like a vertebra from the backbone of a dinosaur. So, I put it in my pocket.' When Farrell returned home with Izzy, he placed the fossil on the kitchen table, where it remained 'for a couple of days'. He then decided to take it to Dinosaur Isle, the island's specialist museum at Sandown. And when he heard the opinion of the experts he declared, 'I was really shocked to find out it could be a new species!'[1]

Said Museum curator, Dr Martin Munt, 'I saw straight away that this was very, very different.' Said Dr Neil Gostling of the University of Southampton, 'It is the most exciting thing because for a moment, you and maybe one or two other people, know something that no one else in the world does.'[2]

Farrell was not the only one to make a discovery. Robin Ward, a regular fossil hunter from Stratford-upon-Avon in Warwickshire, was with his family visiting the Isle of Wight when they found two interesting fossilised bones. Said he, 'The joy of finding the bones was absolutely fantastic. I thought they were special.' And he too, took the fossils to the Dinosaur

Isle museum. The museum authorities 'immediately knew these were something rare and asked if we could donate them to the museum to be fully researched'.[3] This was agreed.

A third visitor also made a significant find. James Lockyer from Spalding, Lincolnshire was also on a visit to the island when he found another fossilised bone. Said he, 'It looked different from marine reptile vertebrae I have come across in the past. I was searching a spot at Shanklin [on the east side of the island] and had been told and read, that I wouldn't find much there. However, I always make sure I search the areas others do not, and on this occasion, it paid off.' Lockyer also donated his find to the Dinosaur Isle museum.

These fossils were discovered on three separate occasions just weeks apart in 2019, and all four fossilised bones were found in the same location, i.e. on the foreshore near Knock Cliff, Shanklin. Furthermore, it was thought likely that they had all belonged to the same individual.[4]

Scientific opinion

Palaeontologists from the University of Southampton were brought in. They opined that 'the creature lived in the Cretaceous period 115 million years ago, and estimated that it had been a creature some 4 metres [13 feet] in length.

'The dinosaur is a new species of theropod. These creatures are all known for being predators, with sharp teeth and claws to help them devour their prey.'[5] It belongs to Teturanae, a group that includes most theropod dinosaurs, including megalosauroids, allosauroids, tyrannosauroids [including Tyrannosaurus rex], ornithomimosaurs, maniraptorans and birds.[6]

Naming the new dinosaur

Said Gostling, 'We bounced the ideas between us. We talked to the people at the museum; we talked to my PhD student (Chris Barker).' The question was, 'what name sounds sensible, because you want the name to tell you something'.[7] The name that was decided upon, *Vectaerovenator inopinatus*, most certainly did – for those who understand Latin, that is! 'Vecta' was the Roman (Latin) name for the Isle of Wight; 'aero' is Latin for 'relating to air'; 'venator' is Latin for 'a hunter'; 'inopinatus' is Latin for unexpected.

Why opinatus? Dinosaurs were not marine creatures. They were terrestrial and tended to live in the vicinity of freshwater lakes. Said Barker,

'You don't usually find dinosaurs in the deposits at Shanklin as they [the deposits] were laid down in a marine habitat.'[8] Therefore, 'it is likely that the *Vectaerovenator* lived in an area just north of where its remains were found, with the carcass having washed out into the shallow sea nearby'.[9]

Why 'aero'? Because there were air cavities in the skeletal vertebrae.

Which vertebrae were they?

They were from the neck (cervical), upper back (thoracic) and tail.

Air spaces in the skeleton

Even a cursory look at the fossilised vertebral bones indicate that it was full of holes. Said Barker, 'We were struck by just how hollow this animal was. It's riddled with air spaces. Parts of its skeleton must have been rather delicate.'[10] But, this is not necessarily the case. One has only to look at a heavy lift crane and how it is constructed to realise this. It is the diagonal bracing of the tower and arm of the crane that give added strength to the structure and ensure that the forces placed upon it are directed in the right manner. Similarly, the trabeculae (bands or columns of bone) adjacent to the air holes would give the vertebrae in question enormous strength. Had the bones been 'delicate', such a large creature would inevitably have collapsed.

The idea that there were pockets of air within the vertebrae may appear fanciful. However, when the skeletons of birds are examined, since birds are the descendants of the dinosaurs, this is much more plausible.

Skeletal air spaces and their advantages

Obviously, anything that increases the capacity of a creature to absorb oxygen from the air and to expel carbon dioxide must be beneficial. 'These air sacs, also seen in modern birds, were extensions of the lungs which helped fuel an efficient breathing system while also making the skeleton lighter.'[11]

In the case of the pigeon, for example, the trachea (windpipe) connects with each of its two lungs by way of bronchi and bronchioles (breathing tubes). But in addition, many large air sacs, roughly equal in volume to that of the lungs, are connected to the lungs to provide extra gaseous exchange in the tissues and organs, thus enabling oxygen to be supplied and carbon dioxide to be vented.

Said Gostling, 'If you are a fast, active hunter with an efficient air [respiratory] system and an ability to take more oxygen out of the atmosphere than just two bags [i.e. the contents of two lungs], then that's going to be a really useful thing; so these holes are an extension of the lungs and allow the animal to be a fast and active predator.'[12]

On 14 August 2020, an article appeared in the *Isle of Wight County Press* by Lori Little, entitled 'New Dinosaur Species found at Shanklin'. It included an image by palaeontologist Darren Naish indicating where in the skeleton the four vertebrae may have been located. Also included was an artistic impression by artist Trudie Wilson of the body of the unfortunate *Vectaerovenator inopinatus* being carried out to sea.

Summary

'The Isle of Wight is famed for its dinosaur remains, with over 25 species having been found on the island.'[13]

Said Dr Chris Barker, 'The record of theropod dinosaurs from the mid-Cretaceous period in Europe isn't that great, so it's been really exciting to be able to increase our understanding of the diversity of dinosaur species from this time.'

Said museum curator, Dr Martin Munt, 'This remarkable discovery of connected fossils by three different individuals and groups, will add to the extensive collection we have and it's great we can now confirm their significance and put them on display for the public to marvel at ... We continue to undertake public field trips from the museum and would encourage anyone who finds unusual fossils to bring them in so we can take a closer look. However, fossil hunters should remember to stick to the foreshore and avoid going near the cliffs, which are among the most unstable on the island.'

Said Isle of Wight Council Cabinet member for environment and heritage, Councillor John Hobart, 'This is yet another terrific fossil find on the island which sheds light on our prehistoric past: all the more so that it is an entirely new species. It will add to the many amazing items on display at the museum.'[14]

Chapter 27

Birds: The Dinosaurs that Did Not Die

This narrative commenced with images of birds, flying high over the waves on the coast of Dorset's Isle of Purbeck, where they build their nests in the cliffs, fish, or sun themselves on the rocky ledges. This mention of birds was no accident, as will now be seen.

Archaeopteryx: An Early Bird-Like Dinosaur

In 1861, not long after the publication of Charles Darwin's groundbreaking work *On the Origin of Species by Means of Natural Selection* (in November 1859), a quarry worker in Bavaria, Germany, discovered an unusual fossil in the Solnhofen (limestone) Formation. It was the impression of a single feather and it dated from about 150 mya (Late Jurassic), when Europe consisted of islands in a shallow, tropical sea close to the Equator. In 1861, German palaeontologist Hermann von Meyer (1801–69) opined that the feather was a 'remnant of a bird from pre-Tertiary times [i.e. pre-KT boundary]', and he named it *Archaeopteryx lithographica* ('ancient wing from the lithographic limestone').

In the summer of 1861, an almost complete skeleton was discovered, together with feather impressions, in the same Solnhofen Formation. However, it was missing most of its head and neck, whereupon von Meyer recognised in it both reptilian and avian (bird-like) characteristics. The specimen subsequently came into the possession of local physician and fossil collector Karl Häberlein, who sold it to London's Natural History Museum for a considerable sum of money.

Sir Richard Owen (1804–92) examined the fossil and pronounced that although it possessed many features reminiscent of reptiles, it was in fact a bird. Included in the definition of a bird is that it is 'typically able to fly'. Nonetheless, there are many birds that are flightless. However, if *Archaeopteryx* could fly, which is not certain, it could indeed be classed as a bird.

As it was not possible to prove that the above specimen and von Meyer's feather were from the same species of *Archaeopteryx*, Owen named the skeleton *Archaeopteryx macrura* ('long-tailed ancient wing').

Biologist Thomas Henry Huxley (1825–95) compared the *Archaeopteryx* skeleton with that of the small theropod dinosaur *Compsognathus*, which was discovered in the same Solnhofen Formation (it is estimated that an adult *Archaeopteryx* was 1 foot 8 inches long and weighed just over 2 pounds. Whereas *Compsognathus* was slightly larger, and about the size of a small chicken). Huxley found that the skeletons had no less than thirty-five characteristics in common.

Sinosauropteryx prima

In August 1996, the fossil of a feathered dinosaur was discovered by Li Yumin, a farmer and fossil hunter, while prospecting in Liaoning Province in Northeast China. *Sinosauropteryx prima* ('Chinese reptilian wing') was a small theropod, 3 feet 6 inches in length and 1.2 pounds in weight, and the fossil dated from 121–135 mya (Early Cretaceous). Since then, many other species of feathered dinosaur have been discovered.

Recent Developments

In July 2014, further light was shed on the subject of feathered dinosaurs, when a paper was published by Pascal Godefroit *et al.* of the Royal Belgian Institute of Natural History, Brussels, Belgium.

Hitherto, all thirty or so species of dinosaur known to have possessed feathers were theropods. Nonetheless, said Godefroit, 'Quill-like structures' had been reported in the ornithischians *Psittacosaurus* and *Tianyulong*, 'but whether these were true feathers, or some other epidermal appendage, is unclear'.[1] (*Psittacosaurus*: a ceratopsian dinosaur of the Early Cretaceous. *Tianyulong*: a genus of heterodontosaurid dinosaur of the Late Jurassic.)

Godefroit *et al.* now reported on the discovery of fossilised dinosaur bones in south-eastern Siberia, Russia, in the locality of Kulinda in the Cherynyshevsky District of Chita Region: 'Six partial skulls and several hundred disarticulated skeletons [were] unearthed from two neighboring monospecific bone beds. Each individual skeletal element is represented by a single morphotype.' (Monospecific: relating to or consisting of only one species. Monotype: member of the same species.[2])

In other words, all the bones referred to belonged to the same species of 'basal ornithischian'.

The dinosaurs to which the bones belonged were described as neornithiscians. The name given to this newly discovered species was

Kulindadromeus zabaikalicus. (Neornithishia: a clade of the dinosaur order Ornithischia and a sister group of the Thyreophora.)

The fossilised bones were discovered in the Ukureyaskaya Formation of rocks, which date from the Bajocian Age (170.3–168.3 mya) of the Mid-Jurassic to the Tithonian Age (152.1–145 mya) of the Late Jurassic.

Kulindadromeus zabaikalicus 'was a small, 1.5-m-long [about 5 feet] bipedal herbivore, with a short skull, plant-eating teeth, elongate hindlimbs, short forelimbs, and an elongate tail'. What was significant was the fact that it possessed 'small scales around the distal hindlimb, larger imbricated scales around the tail, monofilaments around the head and the thorax, and more complex featherlike structures around the humerus, the femur, and the tibia'.[3] (Imbricated: arranged in an overlapping manner like roof tiles.[4])

Conclusion

Said Godefroit *et al.*: 'The discovery of these branched integumentary structures outside theropods [i.e. in a basal, neornithischian dinosaur] suggests that featherlike structures coexisted with scales and were potentially widespread among the entire dinosaur clade; feathers may thus have been present in the earliest dinosaurs.' (Integument: a tough outer protective layer.[5])

Godefroit remarked that 'these animals [i.e. *Kulindadromeus zabaikalicus*] couldn't fly'.[6]

Similarly, it is doubtful whether any of the feathered dinosaurs had sufficient plumage to enable them to fly (with the possible exception of *Archaeopteryx*), and therefore they cannot be classed as birds. However, given that dinosaurs had hollow bones and that birds are their descendants, it seems quite possible that even if they were unable to fly, some species might well have been able to float on water.

The Relationship between Feathers and Scales

Said Danielle Dhouailly of the *Université Joseph Fourier*, Grenoble, France, 'Developmental experiments in modern chickens suggest that avian scales are aborted feathers, an idea that explains why birds have scaly legs.'[7]

Finally, it is interesting to note that the aforementioned feathered dinosaur *Kulindadromeus zabaikalicus* dates from between 170.3 and 145 mya, whereas *Archaeopteryx* dates from about 150 mya. In other words, it is likely that *Kulindadromeus* predates *Archaeopteryx*, even though the two may have been contemporaries.

Genes and the Evolution of Feathers

Said Professor Cheng-Ming Chuong of the University of Southern California, in Los Angeles, 'Modern feathers involve a range of different genes working together and being expressed at the right time and in the right space during the embryo's development.' In fact, there are many dozens of these 'feather-forming' genes. It was likely, he said, that there were 'five separate genetic processes active in birds that needed to work together to create modern feathers'.[8]

Overview

Science writer and author John Pickrell said: 'The oldest feathered dinosaurs we know of so far are around 160 million years old, but if feathers evolved just once and pre-dated the split between ornithischian and saurischian dinosaurs, then they must have first appeared more than 200 million years ago.'

Referring to the work of Dhouailly *et al.*, Pickrell declared: 'The scientists believed *Kulindadromeus* showed that the ability to produce feathers may have evolved early in the history of dinosaurs, perhaps in the Triassic, more than 230 million years ago. These feathers were not used for flights until much later, in the theropod ancestors of birds, and most dinosaurs used them for insulation and [courtship] display.'[9]

The Absence of Feathers in the Larger Sauropods

Commenting on the lack of feathers, as demonstrated in the 'skin impressions in some fossils' of large sauropod dinosaurs, Pickrell opined that 'these feathers may have been lost as animals grew to large sizes, much as the largest land mammals today – including elephants, hippos, and rhinos – have lost the majority of their covering of fur'.[10]

Classification of Birds

It is generally agreed today that 'the gradual evolutionary change – from fast-running, ground-dwelling bipedal theropods to small, winged flying birds – probably started about 160 million years ago'.[11]

Finally, the lineage of birds is likely to be as follows: Saurischia, Theropoda, Tetanurae, Coelurosauria, Manoraptora, to birds.

Chapter 28

How Did Birds Survive the KT Extinction Event?

A case has been made that the dinosaurs became extinct because none of them were hibernating at the time of the KT extinction event or had the capacity to go into hibernation at that point in time. How, therefore, did some species of birds – which are now recognised as theropod dinosaurs – manage to survive, for this would require the capacity to withstand cold, and also starvation for a prolonged period of several weeks or months? Had these avian survivors retained the ability to hibernate, having inherited it from their ancestors, the reptiles? If so, this poses the question, are there any hibernators among today's birds? The answer is yes.

The key to understanding why all the terrestrial dinosaurs became extinct and why only the avian dinosaurs – i.e. the birds – survived lies with a small bird found in North and Central America. The common poorwill (*Phalaenoptilus nuttalii*) is the smallest member of the North American nightjar family. It is named after English-born US ornithologist Thomas Nuttall (1786–1859).

'The common poorwill has gained fame as the first bird species known to hibernate for weeks or even months under natural conditions. One individual was recorded to remain in hibernation for a least 85 days for the 1947 to 1948 season.'[1] These birds 'can enter hibernation in response to environmental stress (lack of food and/or inclement weather). When hibernating, they typically hide away in rocky crevices for several weeks (mostly in winter time) and emerge in spring when the temperature has risen and food (insects) are readily available.'

'Hibernation involves slowing the metabolic rate, dropping the body temperature down dramatically, and a slowing heart rate. This allows a bird to go without food for extended periods and survive cold spells.' It is possible, 'although not officially proven – that other members of the typical Nightjar family [order Caprimulgiformes] also hibernate'.[2]

The nightjar dates from the Mid-Paleocene Epoch (61.6–59.2 mya) to present time. Birds similar to the nightjar and its immediate ancestors would therefore almost certainly have been in existence at the time of the KT boundary and, like its descendant the common poorwill, have possessed the ability to hibernate.

However, there was another way birds could survive, which did not require hibernation. This is exemplified by the penguin.

Gerald Mayr of the Senkenberg Research Institute and Natural History Museum, Frankfurt, Germany, described the partial skeleton of a new species of penguin from the Moeraki Formation at Hampden Beach in the Otago region of South Island, New Zealand. It dated from the Thanetian Age (59.2–56 mya) of the Late Paleocene Epoch, and it was given the name *Kumimanu biceae*. The length of this giant penguin was estimated at 1.77 metres (5 feet 9 inches) and its weight 101 kg (about 220 pounds).[3]

This creature was, therefore, in existence from at least 10 million years after the KT boundary event (of 66 mya). However, by about 20 mya (in the Early Miocene Epoch of the Neogene Period), it had become extinct. As it had existed for almost 40 million years, it is highly likely that it was in existence before the KTB event, and therefore that it had survived this event. But how did it achieve this, when penguins – at least those of the modern day – are not known to hibernate? A clue is provided by the behaviour of modern-day penguins.

Said Yvon Le Maho, Director of the Centre d'Ecologie et Physiologie Energétiques, Strasbourg, France, *et al.*: 'Emperor penguins breed during the cold Antarctic winter. The males incubate the single egg, while fasting for up to 4 months and losing some 20 kg of their body mass [i.e. between 20 per cent and 40 per cent].' In respect of this loss in weight, an estimated 15 per cent was due to water loss, and of the remainder, 95 per cent due to fat metabolism and 5 per cent due to protein metabolism.

Rigidity of the feathers explains why winds of moderate speed, up to 5 m/s [i.e. 11 mph] had little adverse effect in respect of heat loss from the body. Shivering was observed at air temperatures of −8 to −13 degrees Celsius. At very low temperatures the behaviour of huddling close together is essential in reducing metabolic rate. Without this behaviour, survival during the long fast at winter temperatures would be impossible.

Here, it should be mentioned that, when in a huddle, penguins take it in turns to occupy the more exposed positions on the perimeter.[4]

In regard to wakefulness, 'Penguins sleep for a few minutes at a time any time of the day or night, wherever they are, although they sleep for longer periods on land and at night. They will sleep in a wide range of positions –

sitting in the water, standing up, lying down, sitting or in some species, even perched in a tree.'[5]

Summary

It seems likely that birds that survived the KT extinction event fall into two categories: those such as the common poorwill, that had the ability to hibernate (and there were probably other species which had the same ability), and those such as the male emperor penguin, which could survive several months of starvation and keep warm by huddling together with its fellows. However, the penguin is a flightless bird, and therefore not typical of the order, so the vast majority of KT event survivors would be hibernators.

In the case of both the common poorwill and the emperor penguin, the insulation provided by their feathers and down was of the utmost importance.

Chapter 29

The Enduring Attraction of Dinosaurs

In 1987, US science writer Paul S. Taylor stated that, 'Ever since the first dinosaur reconstructions in the mid-1800's, dinosaurs have been big business. They have been used to sell everything from breakfast cereal to gasoline. And now interest is greater than ever. A new craze for dinosaurs and related merchandise is sweeping America and other western nations.

'Almost anywhere children go these days, they are exposed to dinosaurs in one way or another, even on school milk cartons. Even adults are fascinated by these great beasts–and likewise the history and controversy surrounding them.

'Part of the craze has evidently developed in the wake of a series of high-tech dinosaur exhibits currently touring America. Featured at museums and even a major Las Vegas hotel, these large dinosaurs (half-scale) actually move and growl. These traveling 'animated' dinosaurs are attracting huge crowds and have broken long-standing attendance records for several museums.

'The current dinosaur fad is, also, undoubtedly due to:
(a) dramatic new fossil discoveries;
(b) extensive media attention focused on the latest extinction theories;
(c) the widely publicized building of an 18-foot mechanical, flying replica of the great pterodactyl, *Quetzalcoatlus northropi* (now featured in an IMAX-movie shown in museum theaters throughout North America and a PBS-TV documentary).'[1]

In 1993, Canadian palaeontologist, Jordan C. Mallon described his sense of wonder as a child at seeing a film featuring dinosaurs directed by Steven Spielberg. 'It's June 1993 and I just graduated from Grade 5. My dad takes me to see *Jurassic Park* in the theatre as a sort of graduation reward. I'm awed and amazed by the realistic dinosaurs on screen – so much so that I spill my popcorn when the *T. rex* bursts through the electrified

The Enduring Attraction of Dinosaurs

fence. I go home that evening and tell my mom that I'm going to be a palaeontologist when I grow up. Fast forward 22 years later, and *Jurassic World* has just been released in theatres.'[1] This is a film, directed by Juan Antonio Bayona, in which scientists engineer a new breed of dinosaur.[2]

Today, dinosaur memorabilia include mugs and cups, glass figurines, jigsaws, plastic hand-painted toys, and even a hatching dinosaur egg!

And reincarnations of these long-extinct creatures continue to instil a sense of joy and wonder in the minds of adults and children throughout the world.

Epilogue

At the end of the day, what determines whether a species of animal or plant persists is its ability to hand its genetic information (DNA) on to the next generation. Plants do this by producing seeds, spores, roots or rhizomes, which may germinate immediately or lay dormant in the soil for many months – even years. (Spore: a minute, typically one-celled reproductive unit capable of giving rise to a new individual without sexual fusion. Rhizome: a continuously growing horizontal underground stem that puts out lateral shoots and adventitious roots at intervals.[1])

For an animal, however, having reproduced, its offspring is dependent for its survival on its genetic make-up and how well it can adapt to its environment – i.e. it is subject to Darwinian principles.

An animal that has the ability to hibernate, brumate, or enter a state of dormancy/torpor when faced with adverse conditions, such as the cold conditions and scarcity of food that occurred due to the KT extinction event, clearly has an increased chance of survival. And by so doing, it is in effect putting its capacity to reproduce on hold, until conditions become more favourable.

Terrestrial creatures such as mammals and reptiles (including dinosaurs) ultimately depended on plants (fauna) to survive, either directly (in the case of herbivores) or indirectly (in the case of carnivores). The fact that many living creatures survived the extinction of the dinosaurs indicates that the dinosaurs themselves lacked certain inherent biological capabilities that would otherwise have made them 'fit' (i.e. best adapted) to survive the KT event.

Although it is not known whether the asteroid (or asteroids) impacted with Earth during the summer or winter months, it is virtually certain that certain species of terrestrial mammal, reptile, insect, etc., and certain species of freshwater and marine fish were at the time hibernating/brumating or in a dormant/torpid state, as appropriate, either in the northern or southern hemisphere. If not, when faced with this catastrophe and the concurrent prolonged cold and dark conditions, such species would, if possible, have retreated to their burrows, and immediately entered into their respective states of suspended animation.

Epilogue

However, to survive the KT extinction event, it was not sufficient for an animal simply to be able to withstand severe cold. It must also have the ability to hibernate (thereby reducing its metabolic rate), and thus to be able to do without food for a considerable period.

Since the metabolism of the true hibernators would have slowed down considerably, they would have needed neither food nor water. Whereas the brumators may well have accumulated stocks of food, so that when they awoke from time to time, they had something to eat. So, by the time either group came out of hibernation several months after the KT event, the air was beginning to clear (although it would take several years for it to clear completely), sunlight was beginning to break through, and plant life would eventually recover, once the rains had fallen again and washed away or diluted the soot-containing layer. In the meantime, birds and mammals (such as shrews) were able to dig up seeds (which they had stored before the asteroid impact) and exist on those for as long a time as it took for plant life to recover sufficiently and restore the supply.

What of the terrestrial reptiles that survived the KT extinction event by brumating? One might imagine the crocodile, for example, having awoken from its period of brumation, emerging from its burrow and feeding on some of the immense amount of carrion – including dead dinosaurs, which existed following the catastrophe. It could then have re-entered brumation for a further period, by which time outside conditions would have ameliorated to some extent.

Visitors to Swamp Park, Ocean Isle Beach, North Carolina, in January 2019 would have seen brumation in action.

'18 alligators have been frozen in place, as temperatures across several US states plummeted. Images show one of the reptiles frozen solid in the ice, with just his nose sticking out of the frozen water.' This was in order that the creatures could continue to breathe.

George Howard, manager of Swamp Park, posted the incident on Facebook. He said the reptiles had become frozen stuck on Monday night and remained in place all day on Tuesday, as freezing temperatures continued to grip the Carolina states. Mr Howard said that the alligators seem to 'sense' when the water was at freezing point and poked their noses into the air at 'just the right moment'. The park manager explained that the alligators had entered a state of brumation.

Howard told Fox News that 'normally, alligators in the wild "would burrow into the ground" as a form of brumation. But because this particular group lives in captivity at the park, "they have to change the way they're doing it"'.

Furthermore, said Howard, a similar episode had occurred in January 2017, following which 'the alligators thawed out within several days' and did not sustain any injuries 'from being trapped under the ice'.[2]

To say that only those species of Animalia that had the ability to hibernate, brumate, or enter a dormant/torpid state managed to survive the KT extinction event might be an overstatement. However, the fact that so many species that are known to have survived possessed this capacity cannot be ignored.

It is, of course, impossible in the absence of living dinosaurs to prove that these creatures were physiologically deficient in the ability to hibernate, brumate, or become dormant/torpid (with or without the aid of built-in antifreeze). However, as the great fictional detective Sherlock Holmes said, 'when you eliminate the impossible, whatever remains, no matter how improbable, must be the truth'.[3]

Therefore, the lack of this fundamental inbuilt metabolic capability may well have been the reason dinosaurs, the most awesomely powerful creatures that ever walked the Earth, became extinct, as a result of a catastrophic event that occurred 66 million years ago.

A recent event occurred that demonstrates just how vulnerable the Earth would be without its atmosphere, in which most asteroids break up and burn. In January 2019, there took place a total lunar eclipse, producing what has been dubbed a 'super blood wolf moon'.

Said Jorge Zuluaga, astronomer at Universidad de Antioquia, Colombia University, New York City, *et al.*, 'During the total lunar eclipse of 21 January 2019, at least two meteoroids impacted the moon producing visible flash lights on the near side.' (Meteoroid: a rocky or metallic body in the solar system that is significantly smaller than an asteroid.)

One of the impacts occurred on the darkest side of the visible lunar face and was witnessed by many casual observers. The team estimated that the meteorite had impacted with the moon at the speed of 13.8 km ($\pm 4.3/-7.3$ km) per second (about 31,000 mph) at the relatively shallow angle of less than 35.6 degrees.

From 'photometric estimations ... the total impact energy' was equivalent to '0.9 to 1.8 tons of TNT, which corresponds to a body with a mass between 20–100 kg [about 44–200 pounds] and a diameter of 30–50 cm [about 12–20 inches]. If our assumptions are correct, the crater left by the impact will be 7–15 meters [about 7–16 yards] across.'[4]

Epilogue

It is interesting to speculate as to what would happen if a similar event to the asteroid-induced KT extinction event occurred today.

An asteroid that struck the moon 4.3 billion years ago created a crater – the Aitken basin, 1,600 miles or so in diameter – i.e. it impacted about 2 million square miles, or about 1/7th, of the moon's total surface area of 14.6 million square miles.

Given an asteroid of comparable size, whether it impacted the Earth terrestrially or in shallow water (where its energy would be dissipated less than if it landed in the deep ocean), the outcome for Earth's flora and fauna would be similar. However, many species of hibernators, brumators, or creatures that could survive intense cold, darkness, and starvation in a state of dormancy/torpor (such as Greenland sharks) might well survive.

But what of human beings, who, like the dinosaurs, lacked these capabilities? Their survival would depend on whether they could obtain shelter, warmth, and adequate supply of fresh water and food to last for several months until the crisis began to ameliorate.

For the vast majority of the population, this would clearly be an unlikely prospect. Therefore, the human race would be likely to undergo mass extinction, if not total extinction – just like the dinosaurs.

However, there is a faint ray of light at the end of the tunnel. Spacecraft can now track an asteroid and determine if it is on a collision course with Earth. This raises the future possibility of the former being diverted in some way. Also, the possibility of human beings colonising other planets may mean that, one day, mankind will not be dependent solely on the Earth for its survival.

Hibernation and Its Relevance to Space Travel

Because of the vast amount of time that it takes to travel to other planets, it would be advantageous in some ways if space travellers could be induced to enter a state of suspended animation – i.e. hibernation – during the journey. This would considerably reduce their requirement for oxygen and food, for example. Said Mingke Pan: 'Given that most of the genes associated with hibernation are present in the human genome, it could be possible to temporarily induce hibernation using molecular biology techniques.'[5]

What If?

The birds or, in other words, the avian dinosaurs that we know today, survived the KT extinction event, but what if the non-avian dinosaurs had survived? Could they have coexisted with modern man, as we are led to believe in films such as *Jurassic Park*? The smaller species may, by man's condescension, have been permitted to exist in some of the larger forests and savannahs, and even in the Polar Regions. However, the larger species would undoubtedly have shared the same fate as the large modern mammals of today: elephants, giraffes, lions, etc., who find themselves squashed into an ever-decreasing habitat as man overpopulates the planet and denudes and industrialises it to an ever-increasing degree.

An Appreciation of the Birds that are with us Today

Surely, we must be thankful for the birds, the sole representatives of the dinosaurs. Who could fail to appreciate the songs of the blackbird or nightingale; the colourful plumage of finches, parrots, kingfishers; the aerobatic displays of starlings, and the stamina required as they prepare to migrate; the hunting prowess of the eagle; the resilience of the penguin; and the gracefulness of the albatross in flight?

Today, there are an estimated 10,500 species of bird, and countless 'twitchers' (birdwatchers devoted to spotting rare birds[6]) worldwide. How can we fail to appreciate the beauty and prowess of our feathered friends, even those who hunt them for sport (a questionable practice) or depend on them for meat? Birds are, indeed, truly admirable creatures in that they have proved to be incredibly adaptable over the eons of time; so much so that they may even outlive mankind itself.

Appendix

Dinosaur Data

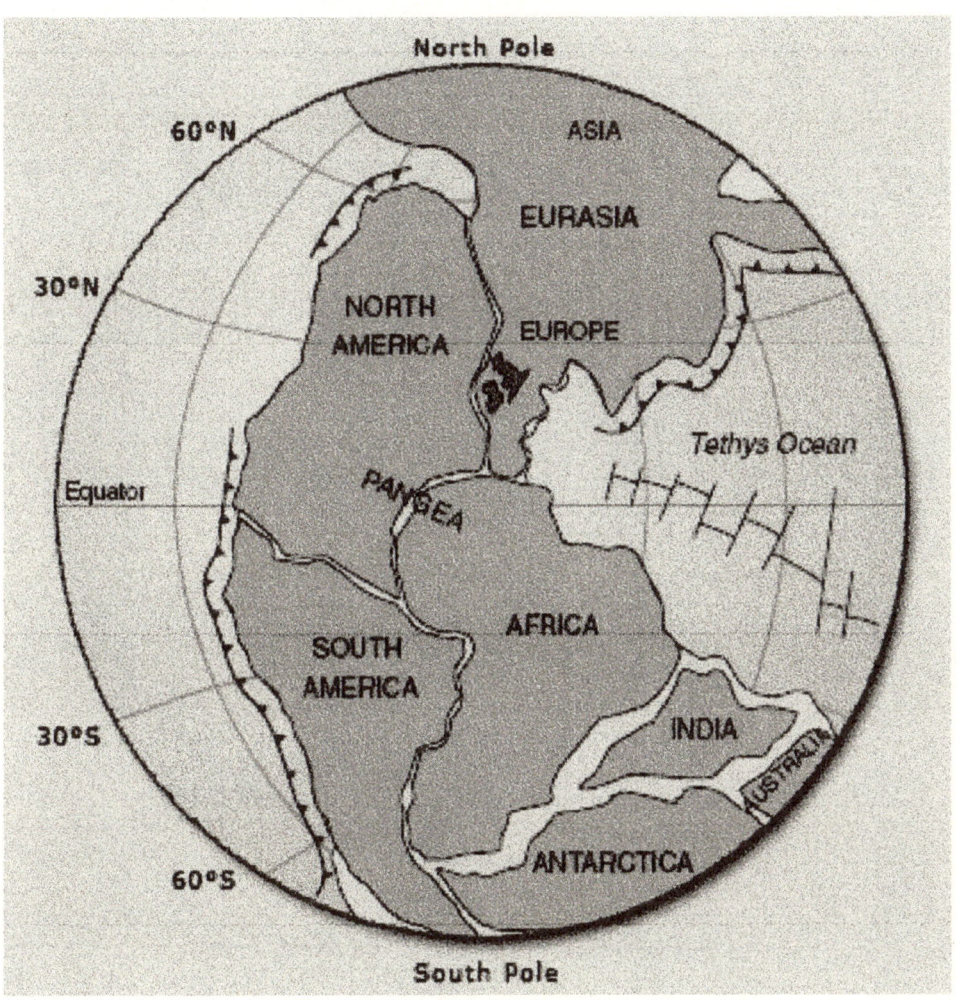

How the continents changed over time: The Triassic Period.
(*Open University*)

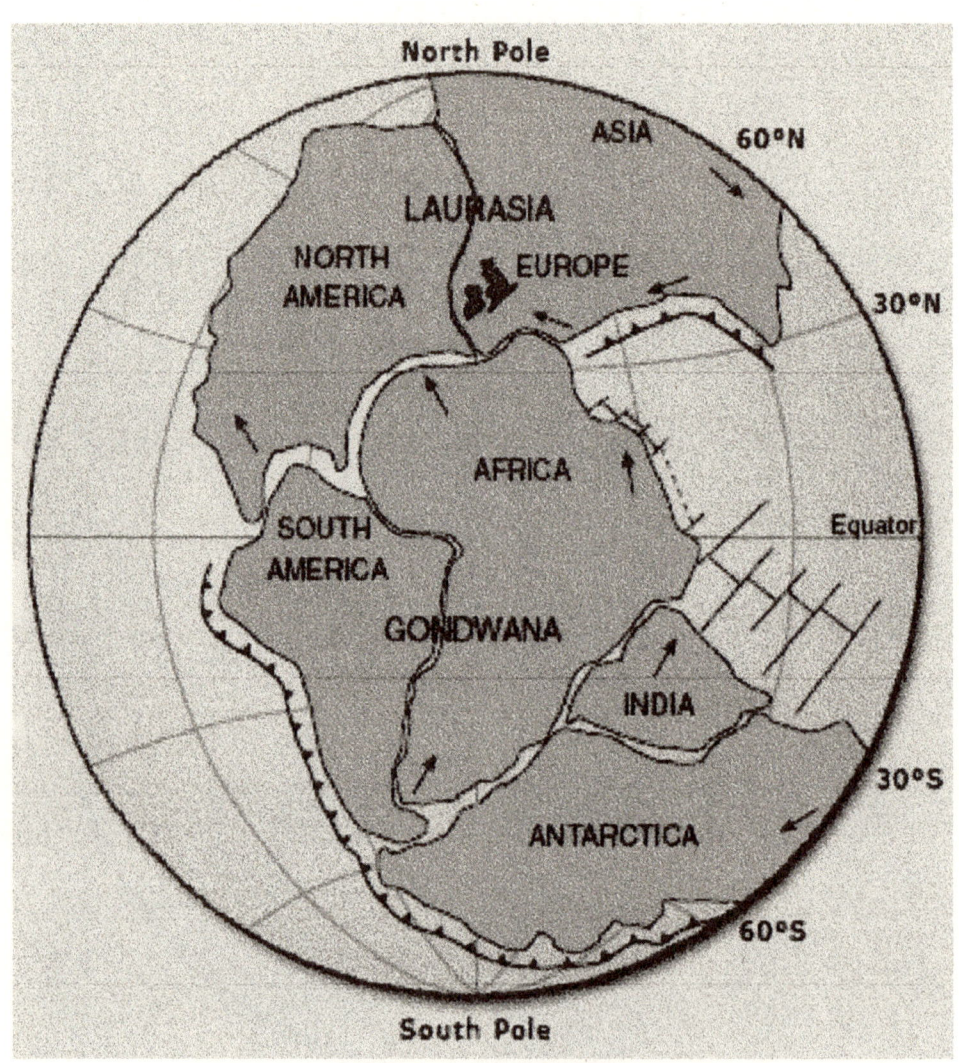

How the continents changed over time: The Jurassic Period.
(*Open University*)

Appendix: Dinosaur Data

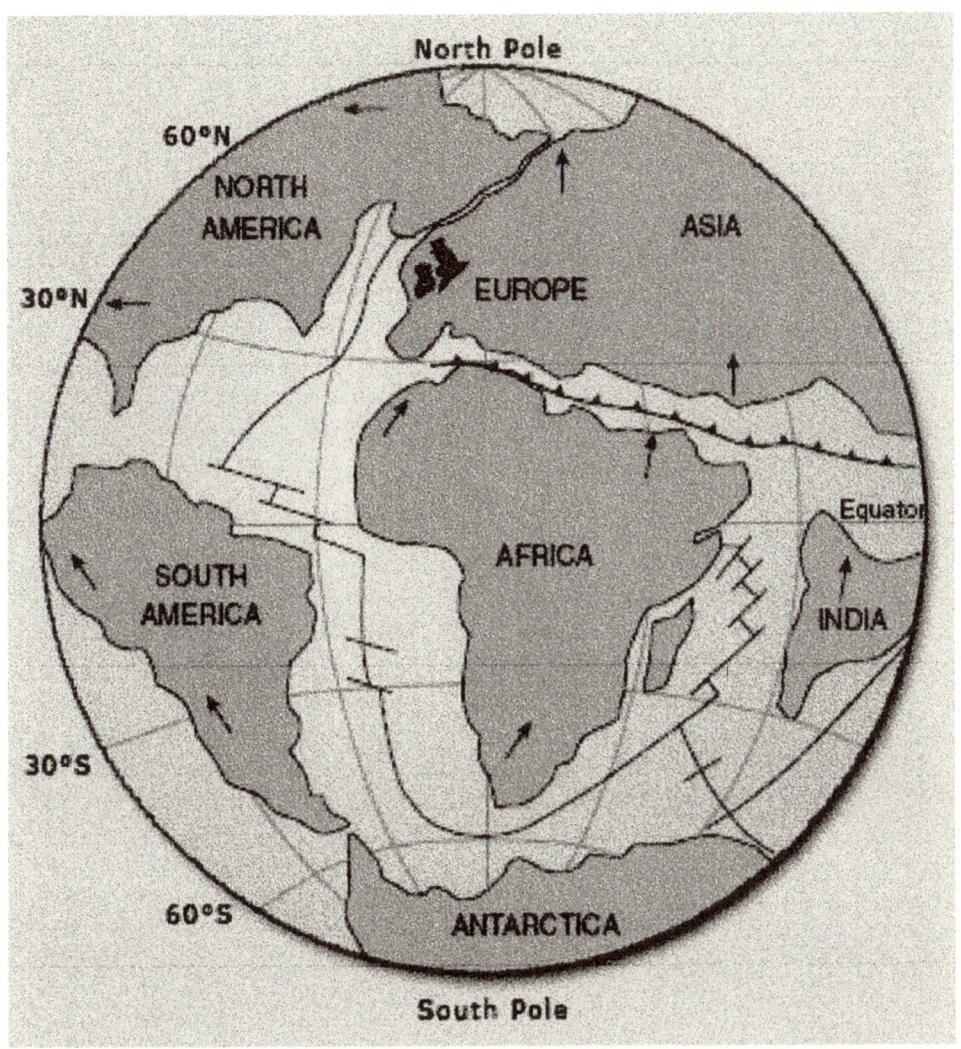

How the continents changed over time: The Late Cretaceous / Early Tertiary Periods.
(*Open University*)

Dinosaur Classification (1) – DINOSAURIA: SAURISCHIA
(Based on *The Encyclopedia of Dinosaurs* by Dougal Dixon)

- **SAURISCHIA** — 'Lizard-hipped'
  - **Sauropodomorpha** — Long-necked herbivores; lizard-like feet
    - **Sauropoda** — Long-necked; elephantine; plant-eaters
      - Vulcanodontidae — Primitive
      - Macronaria — Large nostrils
        - Titanosauria — Late evolving; mainly Southern Hemisphere
          - *Titanosaurus* LC
          - *Duriatitan* LJ
      - Diplodocidae — Long neck; whip tail
        - *Diplodocus* LJ
        - *Apatosaurus* LJ
      - Euhelopodidae — Very long neck
      - Dicraeosauridae — Tall spines on backbone
      - Cetiosauridae — Solid vertebrae
      - Massospondylidae — Long bones in neck
    - **Prosauropoda** — Primitive long-necked herbivores
      - Plateosauridae — Medium-size
      - Melanorosauridae — Large
      - Anchisauridae — Small
  - **Theropoda** — Bipedal carnivores
    - Herrerasauridae — Primitive theropods

Coelophysoidea
  Long and sleek
  Neoceratosauria
    Head crest
    Abelisauria
      Late evolving; mainly Southern Hemisphere

Tetanurae
  Stiff tail
  Coelurosauria
    Lightweight bones
    Compsognathidae
      Small
      Compsognathus LJ–EC
    Therizinosauria
      Large claws on hands
    Troodonidae
      Killing claw on toe
    Ornithomimosauria
      Ostrich-like
    Oviraptorosauria
      Toothless beak
    Alvarezsauria
      Stunted forelimbs
    Deinonychosauria
      Large killing claw on second toe
    Tyrannosauridea
      Late evolving, large
      Tyrannosaurus LC
      Stokesosaurus LJ
    Maniraptora (includes the birds)
  Carnosauria =
    Allosauridae
      Large; carnivores
      Allosaurus LJ
      Metriacanthosaurus MJ
  Spinosauria
    Sail on back; crocodile jaws
    Spinosaurus EC–LC

EJ = Early Jurassic; MJ = Mid-Jurassic; LJ = Late Jurassic; EC = Early Cretaceous; LC = Late Cretaceous.

Dinosaur Classification (2) – DINOSAURIA: ORNITHISCHIA
(Based on *The Encyclopedia of Dinosaurs* by Dougal Dixon)

- **ORNITHISCHIA** Bird-hipped; herbivores
  - **Ornithopoda** Bird-footed; herbivores
    - **Iguanodontidae** Large ornithopods
      - *Iguanodon* EC
      - *Ouenodon* EC
    - **Hadrosauridae** Duck-billed
      - **Hadrosaurinae** Solid crest or none
        - Edmontosaurini — No crest — *Edmontosaurus* LC
        - Maiasaurini — Broad, solid crest above eyes
        - Saurolophini — Pointed crest at top of skull
        - Hadrosaurini — Bulbous nose
      - **Lambeosaurinae** Hollow-crested
  - **Fabrosauridae** Primitive herbivores
  - **Thyreophora** Armoured
    - *Scelidosaurus* EJ
    - **Hypsilophodontidae** Small
    - **Ankylosauria** Armoured back
      - **Nodosauridae** Spikes along sides; narrow snout
      - **Ankylosauridae** Broad snout
        - **Polacanthidae** Spikes on shoulders; shields on hips
        - **Ankylosaurinae** Club at end of tail — *Ankylosaurus* LC
    - **Stegosauria**
      - *Stegosaurus* LJ
      - *Dacentrurus* MJ

Heterodontosaurida
Different-sized teeth
Echinodon EC

Marginocephalia
Horns and shield
around skull

Pachycephalosauria
Dome-headed

Pachycephalosaurini
Horned
Pachycephalosaurus LC

Ceratopsia
Horned

Neoceratopsia
Early bipedal; horned

Ceratopsidae
Large; horned-
headed

Centrosaurinae
Single horn on
nose

Ceratopsinae
Horns over eyes

Chasmosaurini
Very large
neck frill
Triceratops LC

EJ = Early Jurassic; MJ = Mid-Jurassic; LJ = Late Jurassic; EC = Early Cretaceous; LC = Late Cretaceous.

The Phanerozoic Eon
(Dates based on International Commission on Stratigraphy)

Aeon	Epoch (million years ago: mya)
PHANEROZOIC	Mesozoic (251.9–66) Palaeozoic (541–251.9)

Cenozoic, Mesozoic and Palaeozoic Eras
(Dates based on International Commission on Stratigraphy)

Era	Period (mya)
CENOZOIC	Quaternary (2.6 to present) Neogene (23–2.6) Palaeogene (66–23)
MESOZOIC	Cretaceous (145–66) Jurassic (201.3–145) Triassic (251.9–201.3)
PALAEOZOIC	Permian (298.9–251.9) Carboniferous (358.9–298.9) Devonian (419.2–358.9) Silurian (443.8–419.2) Ordovician (485.4–443.8) Cambrian (541–485.4)

Cenozoic Era
(Dates based on International Commission on Stratigraphy)

Era	Period	Epoch	Age (mya)
CENOZOIC	Quaternary	Holocene	Meghalayan (0.0042 to present) Northgrippian (0.0082–0.0042) Greenlandian (0.0117–0.0082)
		Pleistocene	Upper (0.126–0.0117) Middle (0.78–0.126) Calabrian (1.8–0.78) Gelasian (2.58–1.8)
	Neogene	Pliocene	Piacenzian (3.6–2.58) Zanclean (5.33–3.6)
		Miocene	Messinian (7.24–5.33) Tortonian (11.63–7.24) Serravallian (13.82–11.63)

Appendix: Dinosaur Data

		Langhian (15.97–13.82)
		Burdigalian (20.44–15.97)
		Aquitanian (23.03–20.44)
	Oligocene	Chattian (27.82–23.03)
		Rupelian (33.9–27.82)
Palaeogene	Eocene	Priabonian (37.8–33.9)
		Bartonian (41.2–37.8)
		Lutetian (47.8–41.2)
		Ypresian (56–47.8)
	Paleocene	Thanetian (59.2–56)
		Selandian (61.6–59.2)
		Danian (66–61.6)

Cretaceous and Jurassic Periods
(Dates based on International Commission on Stratigraphy)

Period	Age (mya)
CRETACEOUS	Maastrichtian (72.1–66)
	Campanian (83.6–72.1)
	Santonian (86.3–83.6)
	Coniacian (89.8–86.3)
	Turonian (93.9–89.8)
	Cenomanian (100.5–93.9)
	Albian (113–100.5)
	Apatian (125–113)
	Barremian (129.4–125)
	Hauterivian (132.9–129.4)
	Valanginian (139.8–132.9)
	Berriasian (145–139.8)
JURASSIC	Tithonian (152.1–145)
	Kimmeridgian (157.3–152.1)
	Oxfordian (163.5–157.3)
	Callovian (166.1–163.5)
	Bathonian (168.3–166.1)
	Bajocian (170.3–168.3)
	Aalenian (174.1–170.3)
	Toarcian (182.7–174.1)
	Pleinsbachian (190.8–182.7)
	Sinemurian (199.3–190.8)
	Hettangian (201.3–199.3)

The Dinosaurs and their Mysterious Demise

Time of First Arrival

Amphibians
370 mya
Late Devonian Period

Reptiles	**Diapsids** (Sauropsids)	**Archosaurs**	**Dinosaurs**
312 mya	307 mya	250 mya	247 mya
Late Carboniferous Period	Late Carboniferous Period	Early Triassic Period	Early Triassic Period

Crocodilians
83.5 mya
Late-Cretaceous Period

Lizards
199 mya
Early Jurassic Period

Snakes
112–94 mya
Mid-Cretaceous Period

Mammals (Synapsids)
225 mya
Mid-Triassic Period

The Taxonomic Classification of *Tyrannosaurus rex*

Category	Taxon	Contents of Taxon
Kingdom	Animalia	All animals
Phylum	Chordata	Virtually all vertebrates
Class	Reptilia	All reptiles
Clade	Dinosauria	All dinosaurs
Order	Saurischia	All lizard-hipped dinosaurs
Sub-Order	Theropoda	Primarily bipedal carnivores
Family	Tyrannosauroidae	All tyrannosaurs and their close relatives
Genus	Tyrannosaurus	The closest relatives of *Tyrannosaurus rex*
Species	*Tyrannosaurus rex*	The individual species known as *Tyrannosaurus rex*

Appendix: Dinosaur Data

Vertebrate Evolution: Times when lines diverged and new species evolved.
(mya: million years ago)

Vertebrate	Class or Clade	First Evolved
Jawless fish	Agnatha	535 mya
Cartilagenous fish	Chondrichthyes	430 mya
Bony fish	Osteichthyes	420 mya
Amphibians	Amphibia	370 mya
Reptiles	Reptilia	312 mya
From the reptiles, the following evolved:		
Dinosaurs	Dinosauria (including Aves – i.e. birds). Both classed as reptiles	247 mya / 121 mya
Mammals	Mammalia	225 mya

Notes

Introduction
1. Stevenson, A. and Waite, M. (editors), *Concise Oxford Dictionary* (Oxford and New York: Oxford University Press, 2011).
2. Croft, L.R., *The Last Dinosaurs* (Chorley, Lancashire, Elmwood Books, 1982).
3. Stevenson, A. and Waite, M. (editors), *Concise Oxford Dictionary* (Oxford and New York: Oxford University Press, 2011).

Chapter 1. The Coastal Town of Swanage, Dorset
1. Williams, Dr Ann, and Professor G.H. Martin, *Domesday Book: A Complete Translation* (London: Penguin Books, 2002), pp. 225–9.

Chapter 2. Swanage: Ammonites and Evidence of …
1. Stevenson, A. and Waite, M. (editors), *Concise Oxford Dictionary* (Oxford and New York: Oxford University Press, 2011).
2. *Ibid.*
3. *Wikipedia.*
4. *Ibid* (and *Wikipedia*: 'Fossil').
5. *Ibid.*
6. *Ibid.*
7. *Ibid.*

Chapter 3. The Jurassic Coast
1. Stevenson, A. and Waite, M. (editors), *Concise Oxford Dictionary* (Oxford and New York: Oxford University Press, 2011).
2. 'When Did Dinosaurs Live?' (London: Natural History Museum [online], 5 June 2018).
3. McNish, J., and L. Hendry, 'Dinosaur Family Tree Radically Rearranged' (Natural History Museum, News, 22 March 2017), and see M. Baron *et al.*, 'A New Hypothesis of Dinosaur Relationships and Early Dinosaur Evolution' (*Nature*, Volume 10, p. 1038, 23 March 2017).
4. 'When Did Dinosaurs Live?', *op. cit.*
5. 'Geological Processes in the British Isles: A Global View of the Earth's History' (Open University [online]).
6. *Ibid.*
7. 'When Did Dinosaurs Live?', *op. cit.*
8. Stevenson, A. and Waite, M., *op. cit.*
9. *Ibid.*
10. 'Geological Processes in the British Isles: A Global View of the Earth's History', *op. cit.*

Chapter 4. Purbeck: Strata in which Dinosaur Fossils ...

1. Lewer, D. and D. Smale, *Swanage Past* (Chichester, West Sussex, UK: Phillimore, 1994), p. 1.
2. West, Ian, 'Introduction to the Geology of the Wessex Coast with Geological Maps' (Southampton University: Faculty of Natural and Environmental Sciences [online]), p. 7.
3. Geological Survey of England and Wales [online] (Maps 342 and (East) 343).
4. *Wikipedia*.
5. Stevenson, A. and Waite, M. (editors), *Concise Oxford Dictionary* (Oxford and New York: Oxford University Press, 2011).
6. Dates derived from P. Ensom and M. Turnbull, *Geology of the Jurassic Coast* (Wareham, Dorset, Coastal Publishing, 2011).
7. Stevenson, A. and Waite, M., *op. cit.*
8. *Ibid.*
9. *Ibid.*
10. Coram, R., *Prehistoric Dorset: The Story of its Fossils* (Wimborne, Dorset: British Fossils, 1988), p. 16.
11. Stevenson, A. and Waite, M., *op. cit.*
12. *Ibid.*
13. Coram, R., *op. cit.*, pp. 16, 18.
14. Stevenson, A. and Waite, M., *op. cit.*
15. *Ibid.*
16. *Ibid.*
17. *Ibid.*
18. *Ibid.*
19. *Ibid.*

Chapter 5. What is a Dinosaur?

1. Barker, K. (editor), *Proceedings of the Dorset Natural History & Archaeological Society*, P.M. Barrett, and M. Maidment, *Dinosaurs of Dorset: Part III* (Volume 132, 2011), p. 157.
2. Stevenson, A. and Waite, M. (editors), *Concise Oxford Dictionary* (Oxford and New York: Oxford University Press, 2011).
3. *Ibid.*
4. In 1856, Owen was appointed superintendent of the Natural History Department of the British Museum, London, and he was instrumental in the establishment of the separate British Museum of Natural History, of which he became the first director in 1881.
5. 'New Study Shakes the Roots of the Dinosaur Family Tree' (University of Cambridge [online], 22 March 2017).
6. Padid, K., 'How Dinosaurs Evolved into Birds' (Natural History Museum [online], 14 May 2018).
7. Stevenson, A. and Waite, M., *op. cit.*
8. *Ibid.*
9. *Ibid.*
10. *Ibid.*
11. McNish, J., and L. Hendry, 'Dinosaur Family Tree Radically Rearranged' (Natural History Museum: News [online], 22 March 2017), and see Baron, M. *et al.*, 'A New Hypothesis of

Dinosaur Relationships and Early Dinosaur Evolution' (*Nature*, Volume 10, p. 1038, 23 March 2017).
12. Stevenson, A. and Waite, M., *op. cit.*
13. *Ibid.*
14. *Ibid.*
15. *Ibid.*
16. *Ibid.*
17. *Ibid.*
18. *Ibid.*
19. *Ibid.*
20. Hecht, J., 'Fossil Dung reveals Dinosaurs did Graze Grass' (*New Scientist: Daily News*, 17 November 2005), in *Science*, Volume 310, p. 1177.
21. Stevenson, A. and Waite, M., *op. cit.*
22. *Ibid.*
23. Chin, Karen, *et al.*, 'Consumption of Crustaceans by Megaherbivorous Dinosaurs: Dietary Flexibility and Dinosaur Life History Strategies' (Scientific Report No. 7, 2017), Article Number 11163.

Chapter 6. The Crystal Palace Dinosaurs ...

1. Hawkins, B.W., 'On Visual Education as Applied to Geology' (read before the Royal Society of Arts, 27 May 1854, *Journal of the Society of Arts*, Number 78).
2. *Ibid.*
3. Stevenson, A. and Waite, M. (editors), *Concise Oxford Dictionary* (Oxford and New York: Oxford University Press, 2011).
4. McCarthy, S, and M. Gilbert, *The Crystal Palace Dinosaurs: The Story of the World's First Prehistoric Sculptures* (London, Crystal Palace Foundation, 1994), p. 17.
5. *The Crystal Palace Gazette*, November 1853.
6. *The Times*, 2 November 1853.
7. McCarthy, S, and M. Gilbert, *op. cit.*, p. 27.
8. Wilford, J.N., *The Riddle of the Dinosaur* (New York, Knopf, 1985).
9. McCarthy, S, and M. Gilbert, *op. cit.*, p. 30.
10. Sotheby, S.L., 'A letter to Shareholders of the Crystal Palace Company, on the Receipt of the Report for the Committee appointed August 9, 1855, to investigate the affairs of the Company'.
11. Stevenson, A. and Waite, M. (editors), *op. cit.*
12. McCarthy, S, and M. Gilbert, *op. cit.*, p. 67.
13. *Ibid.*, pp. 35–6.
14. *Ibid.*, p. 41.
15. Stevenson, A. and Waite, M. (editors), *op. cit.*
16. Doyle, Sir A. Conan, *The Lost World* (London: Hodder & Stoughton, 1912).

Chapter 7. Mary Anning: Fossil Hunter Extraordinaire!

1. Tickell, C., *Mary Anning of Lyme Regis* (Lyme Regis, Dorset: Lyme Regis Philpot Museum, 1998), p. 10.
2. *Ibid.*, p. 11.
3. *Ibid.*, p. 18.

4. *Wikipedia*.
5. Tickell, C., *op. cit.*, pp. 11–12.
6. *Wikipedia*.
7. Tickell, C., *op. cit.*, p. 10.
8. *Ibid.*, p. 14.
9. *Peeps into an Old Playground: Memories of the Past* (1895), referred to in Lyme Regis Museum.
10. McGowan, C. *The Dragon Seekers* (New York: Perseus, 2001), pp. 203–4.
11. Tickell, C., *op. cit.*, p. 23.

Chapter 8. Dinosaur Prints: How to Identify Them

1. Thulborn, T., *Dinosaur Tracks* (London: Chapman and Hall, 1990), p. 22.
2. Stevenson, A. and Waite, M. (editors), *Concise Oxford Dictionary* (Oxford and New York: Oxford University Press, 2011).
3. Thulborn, T., *op. cit.*, p. 27.
4. *Ibid.*, pp. 76–7.
5. Kuban, G.J., 'An Overview of Dinosaur Tracking' (The TalkOrigins Archive, 2010), originally published in the *M.A.P.S. Digest*, Mid-America Paleontology Society, Rock Island, Illinois, USA, April 1994.
6. Stevenson, A. and Waite, M. (editors), *op. cit.*
7. Glen J. Kuban to the author, 11 February 2019.
8. Thulborn, T., *op. cit.*, p. 106.
9. Glen J. Kuban to the author, 11 February 2019.
10. *Ibid.*
11. *Ibid.*
12. *Ibid.*
13. *Ibid.*
14. *Ibid.*
15. *Ibid.*
16. *Ibid.*
17. Thulborn, T., *op. cit.*, p. 106.
18. *Ibid.*, p. 242.

Chapter 9. Dinosaur Prints: Some Local Discoveries

1. Stevenson, A. and Waite, M. (editors), *Concise Oxford Dictionary* (Oxford and New York: Oxford University Press, 2011).
2. Barker, K. (editor), *Proceedings of the Dorset Natural History & Archaeological Society*, R.B.J. Benson, and P.M. Barrett, *Dinosaurs of Dorset: Part II* (Volume 130, 2009), p. 143.
3. Stevenson, A. and Waite, M., *op. cit.*
4. Wright, J.L., 'Keates' Quarry Dinosaur Footprint Site: Intermarine Member, Purbeck Limestone Group (Berriasian)' (*Proceedings of the Dorset Natural History & Archaeological Society*, 1997, Volume 119), pp. 185–6.
5. Lewer, D., and D. Smale, *Swanage Past* (Chichester, West Sussex, UK: Phillimore, 1994), p. 2.
6. Oppé, E.F., *The Isle of Purbeck: Sunny Spaces and Dinosaur Traces* (published by the author, 1965), pp. 27–9.
7. Calkin, J.B., *Ancient Purbeck* (Dorchester, Dorset: Friary Press, 1981), p. 4.

Notes

8. *Ibid*, p. 4.
9. Delair, Justin B., 'Multiple Dinosaur Trackways from the Isle of Purbeck' (Dorset Natural History & Archaeological Society, Volume 102, 1980).
10. The Dorset County Museum is owned and managed by the Dorset Natural History & Archaeological Society.
11. Information kindly supplied to the author by Paul C. Ensom.
12. West, I., 'Bibliography of the Purbeck Formation: Vertebrates' [online] (2018).
13. *Ibid*.

Chapter 10. Some Dinosaur Fossils from Purbeck …

1. Stevenson, A. and Waite, M. (editors), *Concise Oxford Dictionary* (Oxford and New York: Oxford University Press, 2011).
2. *Ibid*.
3. Coram, R., *Prehistoric Dorset: The Story of its Fossils* (Wimborne, Dorset: British Fossils, 1988), p. 20.
4. *Ibid*., p. 20.
5. *Ibid*., p. 22.
6. Barker, K. (editor), *Proceedings of the Dorset Natural History & Archaeological Society*, P.M. Barrett, R.B.J. Benson, and P. Upchurch, *Dinosaurs of Dorset: Part III* (Volume 131, 2010), p. 117.
7. *Ibid*., pp. 155–6.
8. Brusatte, S.L. and Benson, R.B.J., 'The systematics of Late Jurassic tyrannosauroids [Dinosauria: Theropoda] from Europe and North America' (*Acta Palaeontologica Polonica*, Volume 58:1, 2013), pp. 47–54.
9. Dougal Dixon to the author, 9 February 2019.
10. Barker, K. (editor), *Proceedings of the Dorset Natural History & Archaeological Society*, P.M. Barrett, R.B.J. Benson, and P. Upchurch, *Dinosaurs of Dorset: Part III* (Volume 131, 2010), p. 159.

Chapter 11. How is the Age of the Fossilised Bone …

1. Stevenson, A. and Waite, M. (editors), *Concise Oxford Dictionary* (Oxford and New York: Oxford University Press, 2011).
2. Wilson, T.V., 'How Do Scientists Determine The Age Of Dinosaur Bones?' [online] (Howstuffworks, 2018).

Chapter 12. How Life Began …

1. Stevenson, A. and Waite, M. (editors), *Concise Oxford Dictionary* (Oxford and New York: Oxford University Press, 2011).
2. Darwin to Hooker, 1 February 1871 (Cambridge University, Darwin Correspondence Project, Letter 7471 [online]).
3. Darwin to Wallace, 28 August 1872 (Cambridge University, Darwin Correspondence Project, Letter 8488 [online]).
4. Stevenson, A. and Waite, M., *op. cit*.
5. *Ibid*.
6. Sagan, C., 'Cosmos: Experiment' (YouTube, 6 February 2007).

7. Wallace, A.R., *My Life: A Record of Events and Opinions* (first published in 1905. New York, Elibron Classics, 2005), p. 190.
8. J.D. Hooker and Charles Lyell to the Linnean Society, 3 June 1858 (Burkhardt, F. and S. Smith, editors, *The Correspondence of Charles Darwin*, Cambridge, UK, Cambridge University Press, 1992, Volume 7), pp. 122–3.
9. Darwin, F. (editor), *Autobiography of Charles Darwin* (Cambridge, UK, Icon Books, 2003), pp. 57–8.

Chapter 13. The KT Boundary and the KT Extinction Event

1. University of California, Davis, USA [online] (Winter, 2003).
2. Stevenson, A. and Waite, M. (editors), *Concise Oxford Dictionary* (Oxford and New York: Oxford University Press, 2011).
3. *Ibid.*
4. Chatterjee, S., 'Multiple Impacts at the KT Boundary and the Death of the Dinosaurs' (*Proceedings of the 30th International Geological Congress*, Volume 26, 1997).
5. Chatterjee, S., '225 Million Years of Evolution: The Rise of Birds' (Baltimore, Maryland, USA, Johns Hopkins University Press, 2015), p. 159.
6. *Ibid.*, p. 159.
7. Stevenson, A. and Waite, M., *op. cit.*
8. Chatterjee, S., *op. cit.*, p. 159.
9. Stevenson, A. and Waite, M., *op. cit.*
10. Chatterjee, S., *op. cit.*, p. 160.
11. Stevenson, A. and Waite, M., *op. cit.*
12. *Ibid.*
13. *Ibid.*
14. *Ibid.*
15. *Ibid.*
16. *Ibid.*
17. Brown, M., '*Triceratops* Bones Support Asteroid Extinction Theory', Wired Staff Science [online] (2018).
18. O'Dea, D., Science Educator at the Children's Museum of Connecticut, 'Are Many Fossils Found At The KT Line?' [online] (22 July 2016).
19. 'The Day the Dinosaurs Died' (Barcroft Productions for the BBC, 2017).

Chapter 14. The Presence of High Concentrations ......

1. Alvarez, L.W. *et al.* 'Extraterrestrial Cause for the Cretaceous-Tertiary Extinction' (*Science*, 6 June 1980) Volume 208, Number 4448, pp. 1095–107.
2. *Ibid.*
3. *Ibid.*
4. Stevenson, A. and Waite, M. (editors), *Concise Oxford Dictionary* (Oxford and New York: Oxford University Press, 2011).
5. Alvarez, L.W. *et al.*, *op. cit.*
6. *Ibid.*
7. *Ibid.*
8. *Ibid.*

9. *Ibid.*
10. *Ibid.*
11. *Ibid.*
12. Stevenson, A. and Waite, M., *op. cit.*
13. Alvarez, L.W. *et al.*, *op. cit.*
14. Stevenson, A. and Waite, M., *op. cit.*
15. American Geographical Union, EOS (2 September 1986).
16. Stevenson, A. and Waite, M., *op. cit.*
17. Alvarez, L.W. *et al.*, *op. cit.*
18. Kyte, F.T., 'A Meteorite from the Cretaceous/Tertiary Boundary' (*Nature*, Volume 396, 19 November 1998), pp. 237–9.
19. Stevenson, A. and Waite, M., *op. cit.*
20. Alvarez, L.W. *et al.*, *op. cit.*

Chapter 15. Chicxulub: A Possible Location for ...

1. Chatterjee, S., 'Multiple Impacts at the KT Boundary and the Death of the Dinosaurs' (*Proceedings of the 30th International Geological Congress*, Volume 26, 1997).
2. Stevenson, A. and Waite, M. (editors), *Concise Oxford Dictionary* (Oxford and New York: Oxford University Press, 2011).
3. Renne, P.R., 'Time Scales of Critical Events around the Cretaceous-Palaeogene Boundary' (*Science*, 339 (6120): 684, 8 February 2013).
4. National Aeronautics and Space Administration, Washington D.C., USA (NASA).
5. Braun, David Max, 'Asteroid Terminated Dinosaur Era in a Matter of Days' ('Changing Planet', *National Geographic*, 4 March 2010).
6. Stevenson, A. and Waite, M., *op. cit.*

Chapter 16. Was there more than one Asteroid Impact?

1. Sutherland, F.L. 'The Cretaceous/Tertiary Boundary Impact and its Global Effects with reference to Australia' (AGSO, *Journal of Australian Geology and Geophysics*, 16: 4, 1996), pp. 567–85.
2. Chatterjee, S., 'Multiple Impacts at the KT Boundary and the Death of the Dinosaurs' (*Proceedings of the 30th International Geological Congress*, Volume 26, 1997).
3. Chatterjee, S., '225 Million Years of Evolution: The Rise of Birds' (Baltimore, Maryland, USA, Johns Hopkins University Press, 2015), p. 163.
4. *Ibid.*, p. 169.
5. *Ibid.*, p. 165.
6. Shukla A.D. *et al.*, 'Geochemistry and Magnetostratigraphy of Deccan Flows at Anjar, Kutch' (*Proceedings of the Indian Academy of Sciences, Journal of Earth System Science*, 110: 2, June 2001), pp. 111–32.
7. Chatterjee, S., *op. cit.*, p. 163.
8. *Ibid.*, p. 166.
9. *Ibid.*, pp. 165–6.
10. Stevenson, A. and Waite, M. (editors), *Concise Oxford Dictionary* (Oxford and New York: Oxford University Press, 2011).
11. NASA National Aeronautics and Space Administration, Washington D.C., USA.

Chapter 17. Another Potentially Lethal Ingredient

1. Kaiho, K., Department of Earth Science, Tohoku University, Sendai, Japan *et al.*, 'Global Climate Change Driven by Soot at the K-Pg [KT] Boundary as the cause of the Mass Extinction' (*Scientific Reports*: 6, Article No. 28427, 14 July 2016).
2. Galloway, W.E., Professor Emeritus, Institute for Geophysics, University of Texas at Austin, 'Gulf of Mexico' (*GEOExPRO Magazine*, Volume 6, Number 3, March 2009).
3. *Wikipedia*.
4. Stevenson, A. and Waite, M. (editors), *Concise Oxford Dictionary* (Oxford and New York: Oxford University Press, 2011).
5. 'Chicxulub Impact Event: Regional Effects' (Houston, Texas, USA, Lunar and Planetary Institute [online], 2019).
6. *Wikipedia*: 'Kuwaiti Oil Fires'.
7. Simons, M., 'British Study Disputes Lengthy Climatic Role For Kuwait Oil Fires' (*New York Times*, 16 April 1991).
8. Ross, J. *et al.*, 'Particle and Gas Emissions from an In Situ Burn of Crude Oil on the Ocean' (*Journal of the Air & Waste Management Association*, Volume 46, 1996), pp. 251–9.
9. Barnea, N., 'Health and Safety Aspects of In-situ Burning of Oil' (Seattle, Washington, USA, National Oceanic and Atmospheric Administration [online]).

Chapter 18. The Volcanoes of India's Deccan Region

1. Stevenson, A. and Waite, M. (editors), *Concise Oxford Dictionary* (Oxford and New York: Oxford University Press, 2011).
2. Chatterjee, S., 'Multiple Impacts at the KT Boundary and the Death of the Dinosaurs' (*Proceedings of the 30th International Geological Congress*, Volume 26, 1997).
3. *Ibid.*
4. Stevenson, A. and Waite, M., *op. cit.*
5. *Ibid.*
6. Chatterjee, S., '225 Million Years of Evolution: The Rise of Birds' (Baltimore, Maryland, USA, Johns Hopkins University Press, 2015), p. 172.
7. *Ibid.*, p. 171.
8. *Ibid.*, p. 172.
9. *Ibid.*, p. 172.

Chapter 19. The Chicxulub Impact: Both a Local and ...

1. Berner, R.A., J.M. VandenBrooks, and P.D. Ward, 'Oxygen and Evolution' (*Science*, Volume 316, 27 April 2007).
2. Stevenson, A. and Waite, M. (editors), *Concise Oxford Dictionary* (Oxford and New York: Oxford University Press, 2011).
3. Wolbach, W.S. *et al.*, 'Global Fire at the Cretaceous-Tertiary Boundary' (*Nature*, Volume 334, 25 August 1988), pp. 665–9.
4. Belcher, C.M., 'Geochemical Evidence for Combustion of Hydrocarbons during the K-T Impact Event' (Proceedings of the National Academy of Sciences of the USA, Volume 106: 11, 17 March 2009), pp. 4112–17.
5. Stevenson, A. and Waite, M., *op. cit.*
6. *Ibid.*

Notes

7. *Ibid.*
8. Harvey M.C. et al., 'Combustion of Fossil Organic Matter at the Cretaceous-Paleogene (K-P) Boundary' (*Geology*, Volume 36, 2008) pp. 355–8.
9. Belcher, C.M., *op. cit.*
10. Stevenson, A. and Waite, M., *op. cit.*
11. *Ibid.*
12. *Ibid.*
13. Bardeen, C. *et al.*, 'On Transient Climate Change at the Cretaceous-Palaeogene Boundary due to Atmospheric Soot Injections' (*Proceedings of the National Academy of Sciences of the USA*, 114: 36: 20178980, 2017).
14. Stevenson, A. and Waite, M., *op. cit.*
15. Xarxa Telemàtica Educativa de Catalunya (XTEC) [online] (2018).
16. Stevenson, A. and Waite, M., *op. cit.*
17. Bardeen, C. *et al.*, *op. cit.*
18. Pierazzo, E., *et al.*, 'Chicxulub and Climate: Radiative Perturbations of Impact-Produced S [Sulphur]-Bearing Gases' (*Astrobiology* [online], Volume 3, Number 1, 5 June 2004).
19. Stevenson, A. and Waite, M., *op. cit.*
20. *Ibid.*
21. *Ibid.*
22. Bardeen, C., *et al.*, 'On Transient Climate Change at the Cretaceous-Palaeogene Boundary due to Atmospheric Soot Injections' (*Proceedings of the National Academy of Sciences of the USA*, 114 (36): 20178980, 2017).

Chapter 20. Some Terrestrial, Semi-Aquatic, and ...

1. Cave, J., 'These 7 Animals Survived What Dinosaurs Couldn't' (HUFFPOST *Environment*, 6 December 2017).
2. Stevenson, A. and Waite, M. (editors), *Concise Oxford Dictionary* (Oxford and New York: Oxford University Press, 2011).
3. Bumblebee Conservation Trust [online] (9 October 2014).
4. Stevenson, A. and Waite, M., *op. cit.*
5. *Ibid.*
6. *Ibid.*
7. *Ibid.*
8. *Ibid.*
9. Lovegrove, B.G., *et al.*, 'Mammal Survival at the Cretaceous-Palaeogene Boundary: Metabolic Homeostasis in Prolonged Tropical Hibernation in Tenrecs' (*Proceedings of the Royal Society B*, 22 October 2014).
10. Iacurci, J., 'Shrew-like Ancestors Slept Through Dino Extinction' (*Nature World News*, 23 October 2014).
11. Stevenson, A. and Waite, M., *op. cit.*
12. Horovitz, I., *et al.*, 'Cranial Anatomy of the Earliest Marsupials and the Origin of the Opossums' (PLOS/ONE, 16 December 2009).
13. Verhagen, S., 'Switching off: Hibernation in Australia' (*Australian Geographic*, 7 July 2016).
14. Cave, J., 'These 7 Animals Survived What Dinosaurs Couldn't' (HUFFPOST, Environment, 6 December 2017).
15. Bradford, A., 'Crocodiles: Facts & Pictures (*Live Science*, 25 June 2014).

16. Stevenson, A. and Waite, M., *op. cit.*
17. Bates, M., 'The Creature Feature: 10 Fun Facts about the Greenland Shark' (*Science*, 2 March 2014).
18. Stevenson, A. and Waite, M., *op. cit.*

Chapter 21. Were Dinosaurs Warm-Blooded?

1. Stevenson, A. and Waite, M. (editors), *Concise Oxford Dictionary* (Oxford and New York: Oxford University Press, 2011).
2. *Wikipedia*: 'Physiology of Dinosaurs'.

Chapter 22. The Ability to Hibernate or Brumate ...

1. Pan, M., 'Hibernation Induction in Non-Hibernating Species' (*Bioscience Horizons: The International Journal of Student Research*, Volume 11, 1 January 2018).
2. Stevenson, A. and Waite, M. (editors), *Concise Oxford Dictionary* (Oxford and New York: Oxford University Press, 2011).
3. *Wikipedia*.
4. Pan, M. *op. cit.*
5. Grigg, G.C., 'An Evolutionary Framework for Studies of Hibernation and Short-term Torpor' (ResearchGate [online], January 2011).
6. Gordon C. Grigg to the author, 29 January 2019.
7. Rey, K., 'More that 252 Million Years Ago, Mammal Ancestors became Warm-Blooded to Survive Mass Extinction' (*The Conversation* [online], 18 July 2017).
8. Stevenson, A. and Waite, M., *op. cit.*
9. *Wikipedia*.
10. Meng, J., 'How a Fossil helped to Redraw the Mammalian Family Tree' (*Nature*, Volume 444, p. xv, 14 December 2006).
11. Grabianowski, E., 'How Hibernation Works' [online] (howstuffworks, 2018).
12. Stevenson, A. and Waite, M., *op. cit.*
13. Grabianowski, E., *op. cit.*
14. *Ibid.*
15. Stevenson, A. and Waite, M., *op. cit.*
16. Grabianowski, E., *op. cit.*
17. Pan, M. *op. cit.*
18. Stevenson, A. and Waite, M., *op. cit.*
19. Grabianowski, E., *op. cit.*
20. *Ibid.*
21. Varricchio, D.J., 'First Trace and Body Fossil Evidence of a Burrowing, Denning Dinosaur' (*Proceedings of the Royal Society B: Biological Sciences*, 274 (1616), 7 June 2007), pp. 1361–8.
22. 'New Research Sheds Light on South Pole Dinosaurs' (Montana State University [online], 5 August 2011).

Chapter 23. Hibernation/Brumation/Torpor ...

1. Swain, F., 'The Hibernation Switch' (*Pacific Standard*, 13 June 2014).
2. Stevenson, A. and Waite, M. (editors), *Concise Oxford Dictionary* (Oxford and New York: Oxford University Press, 2011).
3. *Ibid.*

4. Li, X-C., and S-Q Wang, 'Ca2+ Cycling in Heart Cells from Ground Squirrels: Adaptive Strategies for Intracellular Ca2+ Homeostasis' (*PLOS ONE*, Public Library of Science, 2011).
5. Wang, S.Q., *et al.*, 'Adaptive Mechanisms of Intracellular Calcium Homeostasis in Mammalian Hibernators' (*Journal of Experimental Biology*, 205, 2002), pp. 2957–62.
6. Stevenson, A. and Waite, M., *op. cit.*
7. *Ibid.*
8. Labandeira, C.C., *et al.*, National Museum of Natural History, Smithsonian Institute, Washington DC (PNAS, Volume 99, Number 4, 10 February 2002), pp. 2061–6.
9. Stevenson, A. and Waite, M., *op. cit.*
10. 'Fish Antifreeze Proteins' (Alexandria, Virginia, USA, National Science Foundation [online], 1996).

Chapter 24. The Genetics of Hibernation

1. *Wikipedia.*
2. Stevenson, A. and Waite, M. (editors), *Concise Oxford Dictionary* (Oxford and New York: Oxford University Press, 2011).
3. *Ibid.*
4. *Ibid.*
5. *Ibid.*
6. *Wikipedia.*
7. Stevenson, A. and Waite, M., *op. cit.*
8. *Ibid.*
9. *Ibid.*
10. *Ibid.*
11. Malan, A., 'The Origins of Hibernation: A Reappraisal' ('Adaptations to the Cold: Tenth International Hibernation Symposium', 1996), pp. 1–6.
12. Stevenson, A. and Waite, M., *op. cit.*

Chapter 25. For the Mammals Shall Inherit the Earth!

1. 'There Are More Mammals Than We Thought' (Oxford University Press, USA, *EurekAlert!* [online], 6 February 2018).
2. 'David Attenborough's Rise of Animals: Triumph of the Vertebrates: Dawn of the Mammals', (BBC 2, 2013).

Chapter 26. A New Dinosaur is Discovered

1. 'South Today', BBC television, 12 August 2020.
2. 'South Today', BBC television, 12 August 2020.
3. Sky News: Science and Tech News [online], 12 August 2020.
4. 'New Species of Dinosaur Discovered on the Isle of Wight', *Island Echo* [online], 13 August 2020.
5. 'Scientists have discovered a new Species of Dinosaur', CBBC Newsround [online], 12 August 2020.
6. 'New Carnivorous Dinosaur Unearthed in Isle of Wight', SciNews [online], 12 April 2020.
7. 'South Today', BBC television, 12 August 2020.

8. 'New Species of Dinosaur Discovered on the Isle of Wight', *Island Echo* [online], 13 August 2020.
9. 'New Dinosaur related to T rex discovered on Isle of Wight', *The Independent* [online], 13 August 2020.
10. 'New Species of Dinosaur Discovered on the Isle of Wight', *Island Echo* [online], 13 August 2020.
11. 'New Species of Dinosaur Discovered on the Isle of Wight', *Island Echo* [online], 13 August 2020.
12. 'South Today', BBC television, 12 August 2020.
13. 'New Dinosaur related to T rex discovered on Isle of Wight', *The Independent* [online], 13 August 2020.
14. 'New Species of Dinosaur Discovered on the Isle of Wight', *Island Echo* [online], 13 August 2020.

Chapter 27. Birds: The Dinosaurs That Did Not Die

1. Godefroit, Pascal, 'Dinosaur Evolution: A Jurassic Ornithischian Dinosaur from Siberia with both Feathers and Scales' (*Science*: Issue 6195, Volume 345, July 2014), pp. 451–5.
2. Stevenson, A. and Waite, M. (editors), *Concise Oxford Dictionary* (Oxford and New York: Oxford University Press, 2011).
3. Godefroit, Pascal, *op. cit.*, pp. 451–5.
4. Stevenson, A. and Waite, M., *op. cit.*
5. *Ibid.*
6. Vergano, Dan, 'Siberian Discovery Suggests Almost All Dinosaurs Were Feathered' (*National Geographic* [online]), 24 July 2014.
7. 'Newly Discovered Fossils hint that all Dinosaurs may have had Feathers' (*IFLSCIENCE* [online]).
8. Cheng-Ming Chuong, *et al.*, 'Multiple Regulatory Modules Are Required for Scale-to-Feather Conversion' (*Molecular Biology and Evolution*, Volume 35, Issue 2, 1 February 2018, pp. 417–430).
9. Pickrell, John, *Weird Dinosaurs*, (New York, Columbia University Press, 2016).
10. *Ibid.*
11. Pavid, Katie, 'How Dinosaurs Evolved Into Birds' (Natural History Museum [online], London, 14 May 2018).

Chapter 28. How Did Birds Survive the KT Extinction Event?

1. Jaeger, E., 'Further Observations on the Hibernation of the Poor-Will' (Chicago, USA, *The Condor*, 51/3, 1949), pp. 105–9.
2. Beautyofbirds.com [online], 18 February 2019.
3. Mayr, Gerald, 'Nature Communications' (Article Number 8: 1927, 12 December 2013).
4. Le Maho, Y., *et al.*, 'Thermoregulation in Fasting Emperor Penguins under Natural Conditions' (*American Journal of Physiology*, Volume 231, 3 September 1976), pp. 913–22.
5. 'New Zealand: Penguins: Frequently Asked Questions' (penguins.net.nz).

Chapter 29. The Enduring Attraction of Dinosaurs

1. Taylor, Paul S., 'Dinosaur Mania and Our Children', Institute for Creation Research [online], 1 May 1987.

Notes

2. Mallon, Jordan C., 'Dino-mania is back. Thanks, Jurassic World', Fossils, Research, Canadian Museum of Nature, Blog [online], 24 June 2015.

Epilogue

1. Stevenson, Angus and Maurice Waite (editors), *Concise Oxford Dictionary* (Oxford and New York: Oxford University Press, 2011).
2. Gamp, Joe, 'Alligators Freeze in Swamp with Noses above the Ice: Watch Shocking Moment' (*Express* [online]), 25 January 2019.
3. Conan Doyle, Sir A., 'The Sign of Four' (*Lippincott's Monthly Magazine*, Philadelphia, Pennsylvania, USA, 1890).
4. Zuluaga, J.I., *et al.*, 'Location, orbit, and Energy of a Meteoroid impacting the Moon during the Lunar Eclipse of January 21, 2019' (Cornell University [online], 29 January 2019).
5. Pan, M., 'Hibernation Induction in Non-Hibernating Species' (*Bioscience Horizons: The International Journal of Student Research*, Volume 11, 1 January 2018).
6. Stevenson, A. and Waite, M., *op. cit.*

Bibliography

Burkhardt, F. and S. Smith, *The Correspondence of Charles Darwin* (Cambridge: Cambridge University Press, 1992).
Calkin, J.B., *Ancient Purbeck* (Dorchester: Friary Press, 1981).
Chatterjee, S., '225 Million Years of Evolution: The Rise of Birds' (Baltimore, MA: Johns Hopkins University Press, 2015).
Chatterjee, S., 'Multiple Impacts at the KT Boundary and the Death of the Dinosaurs' in *Proceedings of the 30th International Geological Congress*, Vol. 26, 1997.
Coram, Robert, *Prehistoric Dorset: The Story of its Fossils* (Wimborne: British Fossils, 1988).
Croft, L.R., *The Last Dinosaurs* (Chorley: Elmwood Books, 1982).
Darwin, F. (ed.), *Autobiography of Charles Darwin* (Cambridge: Icon Books, 2003).
Doyle, Sir A. Conan, *The Lost World* (London: Hodder & Stoughton, 1912).
Ensom, P. and M. Turnbull, *Geology of the Jurassic Coast* (Wareham: Coastal Publishing, 2011).
'Geological Processes in the British Isles: A Global View of the Earth's History' (Open University [online]).
Lewer, D. and D. Smale, *Swanage Past* (Chichester: Phillimore, 1994).
McCarthy, S. and M. Gilbert, *The Crystal Palace Dinosaurs: The Story of the World's First Prehistoric Sculptures* (London: Crystal Palace Foundation, 1994).
McGowan, C. *The Dragon Seekers* (New York: Perseus, 2001).
Pickrell, John, *Weird Dinosaurs*, (New York: Columbia University Press, 2016).
Stevenson, A. and M. Waite (eds), *Concise Oxford Dictionary* (Oxford: Oxford University Press, 2011).
Thulborn, T., *Dinosaur Tracks* (London: Chapman and Hall, 1990).
Tickell, C., *Mary Anning of Lyme Regis* (Lyme Regis: Lyme Regis Philpot Museum, 1998).
Wallace, A.R., *My Life: A Record of Events and Opinions* (first published in 1905) (New York: Elibron Classics, 2005).
'When Did Dinosaurs Live?' (London: Natural History Museum [online], 5 June 2018).
Williams, Dr A. and Professor G.H. Martin, *Domesday Book: A Complete Translation* (London: Penguin Books, 2002).

Index

Abelisaurs 69
Aitken lunar impact crater (or Aitken basin) 58, 117
Albert, Prince 20, 30
Alvarez, Luis W. 55–61, 73
Anning, Joseph 29–30
Anning, Mary 29–31
Anning, Mary ('Molly'), née Moore 29–30
Anning, Richard 29
Anoplotherium 22
Archaeopteryx lithographica 105
Archaeopteryx macrura 105
Argentinosaurus huinculensis 16
Asaro, Frank 55
Australian Geographic 82
Australian Museum 62

Bardeen, Charles G. 67, 73–77
Barker, Chris 102–104
Barnea, Nir 67
Barrett, Paul M. 46
Beche, Henry de la 29
Belcher, Claire M. 71, 78
Berner, Robert A. 71
Birch, Thomas 30
Bombay High, Mumbai, India 62
Boulogne, Countess of 1
Boynton, William 58
Bradford, Alina 82
Braun, David 59
British Geological Survey 8
British Museum 19
Black Ven near Charmouth, Dorset 44
Buckland, William 13, 29

Camargo, Antonio 58
Calkin, J. Bernard 38–39

Charmouth, Dorset 30, 44
Chatterjee, Sankar 51, 62–63, 68–69
Chesterton, G.K. 80
Chicxulub, Mexico 57–63, 65–66, 70, 72–73, 75, 77–78
Chin, Karen 18
Chuong, Cheng-Ming 108
common poorwill (Phalaenoptilus nuttalii) 109–111
Compsognathus 16, 106
Coram, Robert A. 9–10, 44
Crystal Palace 19, 21, 23
Crystal Palace Company 19, 21

Dacentrurus armatus 45–46
Damon, Robert 8
Dancing Ledge 8
Darwin, Charles Robert 23, 25, 48–50, 105
Deinosuchus 82
Delair, Justin B. 39
DeVries, Arthur ('Art') 95
Dhouailly, Danielle 107–108
Dickens, Charles 22
Dicynodon 22
Dinosaur Isle museum, Isle of Wight 102
Dorset County Museum 40, 43
Dorset Natural History & Archaeolgical Society 37, 43
Doyle, Arthur Conan 23–28
Dryptosaurus 22
Duriavenator 46
Durititan 45
Durlston Bay 1, 40, 43–45
Dyrosaurus 82

Echinodon becklesii 44, 46
Egerton, Philip 30

emperor penguin 110–111
English Channel 1
Enniskillen, Lord 30
Ensom, Paul C. 40
Etches Collection 42–43
Evans, Edgar 1
Exmouth, Devonshire 5

Farrell, Paul 101
Frederick II, Augustus, King of Saxony 30
Fitz Grip, Hugh 1

Galloway, William E. 65
Geological Society 13, 19, 29, 31
Giant Penguin (Kumimanu biceae) 110
Gillingham, Dorset 45
Godefroit, Pascal 106–107
Gondwana 6, 17
Gostling, Neil 101–103
Grabianowski, Ed 89–91
Greenland shark (Somniosus microcephalus) 83–84, 117
Grigg, Gordon C. 87–88

Hadrocodium 99–100
Hadrosaurus 22–23
Hawkins, Benjamin Waterhouse 19–23
Harrison, James 44
Harvey, Mark C. 72
Haysom, Treleven 37
Hecht, Jeff 18
Hildebrand, Alan 58
Hobart, John 104
Holliday, Casey 82
Horovitz, Inés 81
Howard, George 115–116
Huxley, Thomas Henry 106
Hylaeosaurus 13
Hypacrosaurus 40

Iacurci, Jenna 81
Ichthyosaurus 22
Iguanodon 11, 13–14, 19–21, 26–27, 37–40, 42, 46

Iguanodon bernissartensis 42
Iguanodon galvensis 42
Isle of Wight 1, 11, 101–102, 104

Jurassic Coast 2, 4–5
Juratyrant langhami 45–46

Kaiho, Kunio 65
Kimmeridge, Dorset 42, 45
Kuban, Glen J. 33–35
Kulindadromeus zabaikalicus 107–108
Kyte, Frank T. 57

Labandeira, Conrad C. 94
Labyrinthodon 22
Lacovara, Kenneth 54
Langton Matravers 37, 39–40
Lankester, Edwin Ray 24
Laurasia 6, 17
Lee, Cheng Chi 93
Lewer, David 8
Liaoning Province, north-eastern China 106
Linnean Society, London 49–50
Li, Xiao-Chen 93
Lockyer, James 102
Lost World, The 19, 23, 28
Lovegrove, Barry G. 81
Luo, Zhe-Xi 100
Lyme Museum 29
Lyme Regis ('Lyme'), Dorset 29–30

Magnosaurus 46
Maho, Yvon Le 110
Maidment, Susannah C. R. 46
Mayr, Gerald 110
Mantell, Gideon 13, 26, 30
Megaloceros 22
Megalosaurus 13–14, 19, 22, 26, 30, 37, 39–40
Megalosaurus bucklandi 30
Megatherium 22
Metriacanthosaurus 46
Meyer, Hermann von 105
Michel, Helen V. 55

Index

Miller, Stanley L. 48–49
Mosasaurus 22
Munt, Martin 101, 104
Museum of the College of Surgeons 19

Naish, Darren 104
National Aeronautics and Space Administration, USA (NASA) 64
Natural History Museum, London 13, 15, 39, 45–46, 50, 105
Nightjar (order Caprimulgiformes) 109
Nunn, John F. 40
Nuthetes destructor 44
Nuthetes 44, 46
Nuttall, Thomas 109
Nyasasaurus parringtoni 16

Oppé, Ernest 38
Oryctodromeus cubicularis 91
Owendon hoggii 45
Owen, Richard 13–14, 19–21, 30, 44–45, 105

Palaeotherium 22
Pan, Mingke 87, 91, 97, 117
Pangea 5, 7
Penfield, Glen 58
Penge Place, Sydenham Hill, South London 19
Peveril Point 1, 11, 42
Philpot, Thomas 29
Pickrell, John 108
Pierazzo, Elisabetta 75–76
Pinney, Anna 31
Plesiosaurus 22, 29
Plesiosaurus macrocephalus 29
Poole Harbour 1
Princeton University, Princeton, New Jersey, USA 23
Psittacosaurus 106
Pterodactyl 22, 24–27, 112

Renne, Paul R. 58
Rey, Kevin 88
Ricqlès, Armand de 85
Ross, John L. 66

Sagan, Carl 49
Sarcosuchus 82
Scelidosaurus harrisonii 44–46
Science 13, 18, 31, 54, 58–59, 65, 75, 95, 108
Sedgwick, Adam 14
Seeley, Harry Govier 14
Schulte, Peter 59
Scott, Robert Falcon 1
'Shieldcroc' (Aegisuchus) 82
Shukla, Anil D. 63
Sinosauropteryx prima 106
Smale, Dennis 8
Smallmouth Sands, Weymouth, Dorset 45
Solnhofen (limestone) Formation 105–106
Sotheby, Samuel Lee 21
'Square and Compass' 42
St Aldhelm's Quarry 37
Stegosaurus 24, 27, 43
Stokesosaurus 46
Stokesosaurus langhami 46
Strand Magazine 23
Studland Bay 1, 5
Sutherland, Frederick L. 62
Suttle, E.W. 38
Suttle, John 38
Swamp Park, Ocean Isle Beach, North Carolina 115
Swanage Bay 1, 13
Swanage, Dorset 1–4, 8, 11, 37–39, 42–43
Swanage Museum & Heritage Centre 42

Teinolophos trusleri 80
Teleosaurus 22
Tenontosaurus 40
Tenrec ecaudatus 81
Thulborn, Tony 32–34, 36
Tianyulong 106
Times, The 21
Titanosaurs 18, 69
Triadobatrachus massinoti 83
Triceratops 40, 54
Tyrannosaurus rex 14–15

United States Geological Survey   79

Vectaerovenator inopinatus   102–104
Verhagen, Shannon   82

Wallace, Alfred Russel   48–50
Wang, Shi-Qiang   94
Ward, Robin   101
Wealden Beds   11
Wilford, John Noble   21
William I, King   1
Wilson, Tracy V.   47

Wilson, Trudie   104
Wohlschlag, Donald E.   95
Wolbach, Wendy   70, 73, 76–77
Woodward, Holly   92
Worth Matravers   37, 42
Wright, Joanna   37–38

Yucatán Peninsula, Gulf of Mexico   58, 73
Yumin, Li   106

Zuluaga, Jorge   116